Decentralised Energy – a Global Game Changer

Christoph Burger
Antony Froggatt
Catherine Mitchell
Jens Weinmann

Published by
Ubiquity Press Ltd.
Unit 322-323
Whitechapel Technology Centre
75 Whitechapel Road
London E1 1DU

First published 2020

The full text of this book has been peer-reviewed to ensure high academic standards. For full review policies, see http://www.ubiquitypress.com/

Suggested citation:
Burger, C., Froggatt, A., Mitchell, C. and Weinmann, J. 2020. *Decentralised Energy — a Global Game Changer.* London: Ubiquity Press. DOI: https://doi.org/10.5334/bcf. License: CC-BY 4.0

Praise for Decentralised Energy

"In the last few years, renewable energy has broken through the cost barrier. But, if it is to become widely adopted, it has to break through an array of country-specific institutional, technical and political barriers. The strength of Decentralised Energy is that it takes seriously each country's context through a range of country case studies. And yet, it pulls the messages together to give us the common challenges that advocates and promoters of renewable energy and decentralised resources must address to take forward and complete a clean energy transition."
Navroz K. Dubash, Professor, Centre for Policy Research, India

"Everyone knows that renewable energy's time has come. An increasingly important issue relates to decentralised resources, and how to use them most efficiently. Governance frameworks and developing new business models are important for both. This book uniquely takes a global view of these intertwined issues, and is a fascinating read for anyone interested in the acceleration of GHG reduction and in coordination factors for a cost effective energy policy."
Dan Kammen, Professor and Chair, Energy and Resources Group, UC Berkeley, Former Science Envoy, US Department of State

"Accelerating the energy transformation is in all likelihood this generation's most significant challenge to solve with little room for error. The authors write: "The last decade has witnessed the beginning of what is likely to be a fundamental, irreversible transformation of the power and wider energy sectors, […] [which] entails both regulatory incentives as well as entrepreneurial initiatives." This book delivers on the high ambition to compare different models and derive critical success factors: it provides a review of different country archetypes with differing needs on their transition paths; on that basis the authors formulate requirements for decisive, transformative top-down governance; they study start-up success stories and categorise underlying business models; and they place these in a three-phased transformation model, concluding on relevant core competencies and success factors. In its essence the book substantiates the "D3" – decarbonisation, digitalisation, decentralisation – as key drivers for the energy transition through a rich range of top-down and bottom up examples. A relevant, timely, and compelling transition synthesis and precious resource for energy transformation practitioners!"
Christoph Frei, Partner, Emerald Technology Ventures (and former CEO & Sec Gen of World Energy Council)

Contents

Figures

Tables

A note from the authors–editors

The energy system is currently undergoing a fundamental transformation. From fossil to renewable energy, from central power plants to distributed, decentralised generation facilities such as rooftop solar panels or wind parks, from utilities to private residents as producers of energy, and from analogue to digital.

The transformation has been triggered by governments and policy makers, who have provided incentives and the regulatory framework for the changes to happen. It is then shaped and accelerated by private individuals, entrepreneurs, and founders, who seize opportunities that emerge within the new configuration of the system. Both groups of stakeholders must deal with a high level of strategic uncertainty: Which regulatory instruments provide an optimal pathway, reconciling environmental objectives with economic efficiency and system reliability? Which business models of start-ups and founders will succeed, and which core competencies are needed in corporations during the transformation process?

This book is intended to reflect the dual, complementary structure of the transformation – top-down *and* bottom-up – and to provide answers how to deal with strategic uncertainty on both sides. It thus aims to combine the topics of governance and business model innovation. Catherine Mitchell, Exeter University, and Antony Froggatt, Chatham House, are in charge of the top-down governance perspective, while Christoph Burger and Jens Weinmann, both at ESMT Berlin, provide a closer look at business model innovation and the bottom-up perspective.

Chapter 1 of this book introduces the main features driving decentralised renewable energy generation. In Chapter 2, Froggatt and Mitchell analyse energy systems from the governance perspective. They invited country experts to describe the regulatory frameworks and governance of renewable energy and distributed energy resources in selected countries, including Australia, China, Denmark, Germany, India, Italy, as well as parts of the United States. The authors use insights into these exemplary countries to review the impact of these policies and structures on developments in the energy sector and draw conclusions on how to improve the policy framework in different stages of the transformation and varying sociocultural contexts.

Chapter 3 has been written and edited by Burger and Weinmann. They observe the global ecosystem of entrepreneurs aiming to leverage opportunities that emerge with the energy transformation: Which innovations do they push into energy markets? How do they experiment with new business models? The authors have interviewed a sample of key innovators and founders in the field of decentralised energy, whose new ventures aim not to rely on state subsidies. The interviews are complemented by one contribution that has been written by one of the founding members of Mobisol, a start-up operating mainly in Eastern Africa. Burger and Weinmann develop a taxonomy of business models based on the insights and identify six core competencies for corporate stakeholders.

Chapter 4 reunites both governance (top-down) and business model innovation (bottom-up) within a three-phase model of the energy transformation. It summarises the conclusions of Chapters 2 and 3 and aggregates core competencies relevant for policy makers, corporate players, and start-ups to deal with strategic uncertainty in a future energy system where all three phases of the energy transformation might co-exist. The book closes in Chapter 5 with a wrap-up and outlook.

The book has been written for policy makers, investors, executives of utilities and corporations, entrepreneurs, as well as a broader academic audience. Especially for readers in developing countries and emerging economies, digitalisation and the internet have made access to scientific literature easier than ever before. However, a major remaining hurdle is the cost to legally acquire digital contents. The authors have therefore chosen Ubiquity as a publisher that follows an explicit open access approach of the dissemination of research.

The authors are greatly indebted to the individuals who volunteered to contribute to this book with country reports: Ranjit Bharvirkar, Søren Djørup, Michele Gaspari, Frede Hvelplund, Arturo Lorenzoni, Dörte Ohlhorst, Liao Maolin, Helen Poulter, Wei Shen, Zhou Weiduo, as well as Klara Lindner for the description of the start-up Mobisol in Chapter 3. Burger and Weinmann would particularly express their gratitude to the founders and entrepreneurs Reza Alaghehband, Jemma Green, Sebastian Groh, Lars Krückeberg, Timo Leukefeld, and Oliver Stahl, who volunteered with their time – the most precious resource that a founder may have – to narrate their stories.

Empirical research, both quantitative and qualitative such as this book, typically suffers from a positive selection bias. Successful examples – countries as well as start-ups – are more likely to be portrayed than flawed policy experiments and founders who failed to achieve lasting commercial success with their ventures and went out of business. The authors, also in their function as editors, tried to avoid neglecting these sobering experiences in the narratives of countries and start-ups.

November 2019,
Christoph Burger, Antony Froggatt, Catherine Mitchell, and Jens Weinmann

This work was in part supported by The Engineering and Physical Sciences Research Council (EPSRC) [EP/N014170/1]

Glossary

AB	State Assembly Bill (California)
AC	Alternating current
ACCC	Australian Competition and Consumer Commission (Australia)
ACT	Australian Capital Territory (region of Australia)
AEMC	Australian Energy Market Commission (Australia)
AEMO	Australian Energy Market Operator (Australia)
AER	Australian Energy Regulator (Australia)
AI	Artificial intelligence
ARENA	Australian Renewable Energy Association
B2B	Business-to-business
B2C	Business-to-customer
BMVBS	Federal Ministry for Construction, Traffic and Urban Development (Germany)
BNEF	Bloomberg New Energy Finance
BoP	Base of the pyramid
BUND	Federation for Environment and Nature Conservation Germany
CA	California (USA)
CAGR	Compound annual growth rate

CAISO	Californian Independent System Operator
CAPEX	Capital expenditure
CARB	Californian Air Resources Board
CCA	Community Choice Aggregation (California)
CCC	Committee on Climate Change (United Kingdom)
CCGT	Combined cycle gas thermal power plant
CCS	Carbon capture and storage
CDU	Christilich-Demokratische Union (Germany)
CEA	Central Electricity Authority (India)
CEC	California Energy Commission
CEO	Chief Executive Officer
CET	Clean Energy Target (Australia)
CFO	Chief Financial Officer
CHP	Combined heat and power
CO2	Carbon dioxide
COAG	Council of Australian Governments (Australia)
CPUC	California Public Utilities Commission
CRM	Customer relationship management
CSU	Christlich-Sozialistische Union (Germany)
CT	Current transformers
CTO	Chief Technical Officer
CXO	Chief X Officer, including chief executive officer, chief financial officer, etc.
D4	Decentralisation; decarbonisation; digitalisation; democratisation
DC	Direct current
DER	Distributed energy resources
DG	Distributed generation
DGRV	Deutscher Genossenschafts- und Raiffeisenverband (German Cooperative and Raiffeisen Confederation)
DNO	Distribution network operator
DNSP	Distribution network service provider (Australia)
DPS	Department of Public Service (New York State)
DR	Demand response
DRP	Distribution Resource Plan (California)
DSIP	Distribution system implementation plan (New York State)

DSO	Distribution system operator
DSP	Distribution system provider
EAM	Earning Adjustment Mechanism (New York State)
ECA	Energy Consumers Australia (not-for-profit advocacy)
EEG	Renewable Energy Sources Act (Erneuerbare Energien-Gesetz)
ENA	Energy Networks Australia
ESB	Energy Security Board (Australia)
ESCO	Energy supply company
ESMI	Prayas Energy Group's Electricity Supply Monitoring Initiative (India)
ESP	Electric service provider
ETS	Emissions Trading System (European Union)
EU	European Union
EV	Electric vehicle
FCAS	Frequency Control Ancillary Services (Australia)
FERC	Federal Energy Regulatory Commission (USA)
FiT	Feed-in-Tariffs
FYP	Five Year Plan (China)
GC	Green Certificates (Italy)
GDP	Gross Domestic Product
GHG	Greenhouse gas
GME	Gestore dei Mercati Energetici SpA (Italy)
GPT	General purpose technology
GSE	Gestore dei Servizi Energetici (Italy)
GSM	Global System for Mobile Communications (Groupe Spécial Mobile)
GW	Gigawatt
GWh	Gigawatt hours
HVAC	Heating, ventilation, and air conditioning
ICO	Initial coin offering
ICT	Information and communication technology
IDCOL	Infrastructure Development Company Limited (Bangladesh)
IEA	International Energy Agency (Paris, France)
IEPR	Integrated Energy Policy Report (California)
IETA	International Emissions Trading Association

IOU	Investor-owned utility
IPEX	Italian electricity market
IRENA	International Renewable Energy Agency (located in Abu Dhabi, United Arab Emirates)
IT	Information technology
KfW	German Development Bank (Kreditanstalt für Wiederaufbau)
kW	kilowatt
kWh	kilowatt hour
LCOE	Levelised cost of electricity
LED	Light-emitting diode
LMP	Locational marginal price (of energy)
LNG	Liquefied natural gas
MBA	Master of Business Administration
MDPT	Market Design and Platform Technology Report (New York State)
MVA	Megavolt Ampère
MW	Megawatt
MWh	Megawatt hours
NEA	National Energy Administration (China)
NECP	National Energy and Climate Plan (Italy)
NEG	National Energy Guarantee (Australia)
NEM	Net energy metering
NER	National Electricity Rules (Australia)
NGO	Non-governmental organisation
NRA	National Regulatory Authority (Italy)
NRDC	National Development and Reform Commission (China)
NSW	New South Wales (region of Australia)
NTEM	Northern Territory Electricity Market (Australia)
NWIS	North Western Interconnected System (Australia)
NY PSC	New York Public Service Commission
NYISO	New York Independent System Operator
NYS	New York State (USA)
OECD	Organization for Economic Co-operation and Development (located in Paris, France)
P2P	Peer-to-peer
PBR	Performance-based regulation

PC	Personal computer
PO	Partner organisation (Bangladesh)
PSO	Public service obligation
PSR	Platform service revenue (New York State)
PURPA	Public Utility Regulatory Policies Act (USA)
PV	Photovoltaic
QLD	Queensland (region of Australia)
R&D	Research and development
RAB	Regulated asset base (Australia)
RE	Renewable energy
RES	Renewable Energy Sources
REV	Reforming the Energy Vision (New York State)
REZ	Renewable Energy Zone (Australia)
RPS	Renewable Portfolio Standard
SA	South Australia (region of Australia)
SB	Senate Bill (California)
SCER	Standing Council on Energy and Resources (Australia)
SEB	State Electricity Board (India)
SEPAP	Solar Energy for Poverty Alleviation Programme (China)
SEU	Sistemi Efficienti di Utenza (Italy)
SME	Small and medium-sized enterprise
SPD	Sozialdemokratische Partei (Germany)
sq	Square
SRAS	Spinning Reserve Ancillary Services (Australia)
SRES	Small-scale Renewable Energy Scheme (Australia)
TAS	Tasmania (region of Australia)
TNSP	Transmission network service providers (Australia)
toe	Tonnes of oil equivalent
TSO	Transmission system operator
TV	Television
TWh	Terawatt hours
UAE	United Arab Emirates
UK	United Kingdom
US/USA	United States of America

UVA	Unità Virtuali Abilitate (Italy)
VIC	Victoria (region of Australia)
VPP	Virtual Power Plant
WEM	Western Electricity Market (Australia)
WiFi	Wireless Fidelity
Wp	Watt peak

CHAPTER 1

Introduction – what are the drivers of decentralised renewable energy generation?

The entire world is moving towards decentralised energy generation. A couple of years ago, this statement would have been the vision of a distant future. No one would have believed that within less than a decade the existing configuration of power supply would be fundamentally challenged. In large parts of the world, decentralised energy generation means *renewable* energy generation, because solar photovoltaic panels and wind turbines are scattered across residential rooftops and and dispersed on acres and farmland. They constitute a fundamental reversal of the paradigm of economies of scale that used to dominate the economics of the energy supply industry in the 20th century.

Of course, the old system of large thermal and nuclear power plants, centralised dispatch, and long-distance transmission lines will co-exist for several decades to come. It brought nation states a reliable supply structure, even though future generations may have to bear the welfare losses for its legacy with respect to climate change, nuclear waste, and stranded assets.

Curbing greenhouse gas emissions has become a global imperative to prevent a lasting, negative impact on the development path of future generations. One of the least contested policy options is a carbon-neutral energy supply. Regulators and politicians have significantly contributed to the rise of renewable generation that promotes a shift towards a sustainable supply structure. In many industrialised countries, they opted for generous subsidy schemes that helped manufacturers of renewable generation technologies, in particular solar and wind, to scale their operations and drive costs down. Now politicians have to find solutions about how to maintain a resilient system in spite of a substantial share of intermittent, weather-dependent, and decentralised renewable supply.

How to cite this book chapter:
Burger, C., Froggatt, A., Mitchell, C. and Weinmann, J. 2020. *Decentralised Energy — a Global Game Changer.* Pp. 1–19. London: Ubiquity Press. DOI: https://doi.org/10.5334/bcf.a. License: CC-BY 4.0

Many emerging economies, most notably China, whose domestic energy policy is discussed in Section 2.3 of this book, but also highly industrialised states such as California, which we explore in Section 2.8, experienced government regulation that has led to a *centralised* dissemination of renewable energies, with large-scale, utility-owned installations of photovoltaic fields and wind parks. Often, this is a fast and efficient way of reducing the carbon footprint of energy supply. However, we believe that the true revolutionary potential of these recent changes of the supply structure relates to the empowerment of the final consumer to transcend into a local, sometimes even autonomous producer of energy, as it occurred in countries such as Australia, Germany, or Denmark.

Most importantly, in developing countries decentralised renewable generation may lead to leapfrogging of certain stages of infrastructure development, analogous to the usage of cell phones instead of building a fixed network for landline telephony services. Especially in rural areas, it may provide a complementary service to the existing energy infrastructure, with individual households establishing micro-grids that enhance commercial activities and, literally, improve the quality of life of local residents.

In industrialised countries, the looming age of decentralised generation does not mean that all utilities will disappear within the next decade. But those utilities that are unable to adapt to the new market environment may one day be swallowed by players from the information and communication technologies or manufacturing sectors, or shrink in their position from providers of a critical infrastructure service to the equivalent of a telephony retailer or private insurance company. A whole range of new players will enter energy markets and redefine business models, revenue streams, and risk allocation. Information and communication providers as well as start-ups occupy commercial niches in decentralised energy generation that utilities are not capable or willing to enter; they provide financing options, technical advice, operation and maintenance of assets, and care for their customers' needs.

Most importantly, though, this movement is not only a global transformation, it is an individual transformation too: across the globe, private consumers decide to turn into micro-investors for their personal generation and, increasingly, storage devices. Collectively they contribute to the renewal and reconfiguration of the energy system.

In the remainder of this first chapter, we want to highlight six key trends that characterise and shape the momentum of change within the energy sector, namely the competitiveness of renewables and decentralised generation (Sections 1.1 and 1.2), the rising role of storage (Section 1.3), the decoupling of growth and energy intensity (Section 1.4), enhancing local value creation (Section 1.5), and digitalisation as enabler of smart grids and new business models (Section 1.6). We will focus on electricity as the segment of the energy sector that is most fundamentally transformed. An analysis of other changes, in particular the role of efficiency and the electrification of transport, would generate equally relevant insights, but is unfortunately beyond the scope of this research.

The introduction will then serve as the basis for the discussion on diverging regulatory models of the electricity supply industry, which will be discussed in Chapter 2, and the emergence of new business models in the context of developing and industrialised countries in Chapter 3. We will conclude the book with a chapter on concrete policy recommendations and the main attributes of successful business models in Chapter 4, which describes the three stages of current, decentralised energy supply, and Chapter 5 as an executive summary of the major findings of both top-down and bottom-up approaches to promoting renewable energies on a global scale.

With this endeavour, our desire is to encourage political and corporate decision makers to assess the most appropriate model to support a future electricity system, adapted to local market conditions, encouraging entrepreneurial activity that minimises the carbon footprint, while ensuring that the conflicting energy triangle of security of supply, resource efficiency, and sustainability is secured for the generations to come.

1.1 Renewables becoming competitive

We may perceive the recent rise of consumer empowerment as a more fundamental disruption than previous changes. Energy policy has undergone major shifts in priorities since the beginning of the 20th century, and each change was perceived as a radical break with the status quo.

Prior to the 1970s, energy policy was primarily focused on affordability; increasing the proportion of the population which had access to energy networks and adding capacity to match economic development. The oil crises of the 1970s led to new directions for energy policy in many parts of the world, including Europe, the US, and Japan. These new energy policies can be broadly divided between those countries or states which tried to improve energy security through reducing oil use by developing other sources of primary energy supply, particularly for electricity, and by using energy more efficiently (for example, California, Denmark, and Japan); and those countries which by and large continued to be dependent on oil – either by developing their own oil resources or attempting to diversify their supply.

Energy policies around the world continued to be dominated by security concerns until environmental matters – including acid rain, the ozone layer, climate change, and local air pollution – began to gain importance in the 1980s and 1990s. In December 1997, the Kyoto Protocol was the first global effort to curb greenhouse gas emissions. Simultaneously, the liberalisation and privatisation of many public infrastructure services also affected regulation in the energy sector. With a focus on increasing efficiency via market mechanisms, liberalisation paved the way to implement competition in the generation and retail segments of the electricity supply industry.

The combination of emission reduction targets with a competitive, market-oriented regulation of the electricity sector has led to an unprecedented rise of renewable energies. This both reduces the dependency on fossil fuels and accelerates the deployment of decentralised, climate-friendly energy sources. If humankind wants to curb carbon dioxide emissions to ensure that global temperature rises remain well below 2°C and strive towards a rise of 'only' 1.5°C, as stated in the Paris Agreement in 2015 and ratified by almost 180 countries, as of August 2018, decentralisation is increasingly seen as a 'no regrets' strategy for meeting the core energy policy goals.

For governments, there is no single trajectory, no 'one size fits all' strategy. A few countries choose nuclear as an (almost) carbon-neutral power generation technology, while in most other countries the nuclear power fleet faces decommissioning within the next decade or two. Cost overruns and severe delays in the majority of new nuclear plant constructions in Europe (Schneider & Froggatt 2019) makes it seem unlikely that the technology will experience a renaissance in the Western world. Similarly, carbon capture and storage (CCS) as a means to reduce greenhouse gas emissions faces severe opposition from local residents, and many pilot projects in the Western world have been prematurely ceased. As these two options do not seem to be politically desired and economically feasible in multiple jurisdictions, it is renewable energy that is the most likely substitute for fossil fuels.

The price of renewable electricity technologies, such as onshore and offshore wind and solar photovoltaics, has fallen rapidly in the last decade. This is because of lower prices due to increased competition, a shift in production to lower-wage economies (from Europe to Asia), technology improvements, and economies of scale. In Europe, the cost of solar modules decreased by 83 per cent between 2010 and 2017. According to an International Renewable Energy Agency (IRENA) estimate, the global weighted average LCOE of utility-scale PV plants has fallen by 74 percent between 2010 and 2018, from US$3,300–7,900 per kW range in 2010 to US$800–2,700 per kW in 2018. The utility scale solar PV projects commissioned in 2018 had a global weighted-average LCOE of US$0.085 per kWh, which was around 13 percent lower than the equivalent figure for 2017 (IRENA 2019).

While even for the more mature wind turbine industry, costs have fallen. For wind in 2018, new capacity was commissioned at a global weighted average LCOE of US$0.056 per kWh, which was 13 percent lower than the value for 2017 and 35% lower than in 2010, when it was USD 0.085 per kWh (IRENA 2019). These falling technology costs and ongoing policy support have led to renewables now dominating new build in the power sector. In the global electricity supply, an additional 181 GW of new renewables capacity was installed in 2018, the largest ever annual increase, 65% of all new supply investment (Ren21 2018). Going forward the trend is expected to continue with solar and wind, according to Bloomberg New Energy Finance, to attract 73 per cent of investment in the power sector between 2017 and 2040 (Henbest 2017). As a consequence, onshore wind and solar PV power are now, frequently, less expensive than any fossil-fuel option, without financial assistance.

Table 1: Examples of Cost of Solar and Wind Projects 2017–9.

Country	Price-US$/ MWh		Details
2017			
India	37	Solar	Indian developer ACME Solar emerged as the winning bidder for a 200 MW project with a tender price of ⍰2.44 per kWh.
Germany	63	Wind	The introduction of auctioning for wind for the first time for onshore wind installations led to an average bid of 5.71 cents per kWh for 70 bids with a total installed capacity of 700 MW – 93% of the bids (65) or 96% of the volume of bids were awarded to citizens' energy companies.
Chile	32.5	Solar/ Wind	An average price of $US32.5 per MWh was awarded for 600MW of solar and wind capacity.
India	38	Wind	A wind energy auction for 500 MW of capacity and organized by the state government of Gujarat revealed a tariff of Rs 2.43 per kWh as the lowest bid.
2018			
Germany	55	Onshore wind	In total 83 bids were awarded for a total of over 700MW of capacity. The range of successful bidders was €38 per MWh to €52.80 per MWh with an average of €47.3 per MWh.
United States	36	Onshore wind	The levelized cost of wind also hit an all-time low, averaging $36 per MWh for plants built in 2018 across the United States.
Brazil	21	Wind	In April 114MW of wind was contracted in a tender at a price of R$67.6 per MWh for capacity contracted from four projects in the north-eastern state of Bahia.
India	35	Solar	In July 2018, ACME Solar quoted the lowest tariff of ⍰2.44 per kWh for 600 MW of solar projects in the Solar Energy Corporation of India (SECI)'s 2 GW ISTS Phase I auction.
2019			
Saudi Arabia	17	Solar	Saudi Arabia's Acwa Power submitted a tariff of just US$c 1.6953 per kW for the 900MW fifth phase of Dubai's Mohammed bin Rashid Al Maktoum (MBR) Solar Park

Table 1: Continued.

Country	Price-US$/ MWh		Details
US	19.9	Solar	In June a 400MW project in Los Angeles was agreed at US$c1.997 per kWh.
Portugal	16.6	Solar	In July, the Direcção-Geral de Energia e Geologia awarded a series of contracts to provide 1.15GW of solar energy. Within that, 150MW was secured for a price of just €0.01476 per kWh.
Saudi Arabia	20	Wind	In August it was announced that the costs of electricity from the 400MW Dumat Al Jandal onshore wind farm would be US$c 1.99 per kWh.
UK	50.0	Offshore wind	12 projects, including 5.5 GW of offshore wind projects, at record low prices as low as £39.65 were contracted.

A broader experience in the siting of renewables, faster installations, and lower related costs, as well as an increase in conversion efficiencies have contributed to further reduce the cost of energy produced from renewables. Significant improvements have been achieved because of a move to renewable auctions, although there are critical voices as to the long-term viability of these cost reductions and their impact on the diversity of market actors. (Klessmann & Tiedemann 2017). Nonetheless, these have resulted in a decrease in the price per megawatt and contracts for large-scale renewable technologies, as can be seen in Table 1.

Although globally renewables are still a relatively small share of total power production, around 5 per cent, in selected countries and regions they have become significant providers of electricity. In 2018, wind energy provided an estimated 11.8 per cent of EU annual electricity consumption – including Denmark, which met 41 per cent of its annual electricity consumption with renewables. Globally, at least 12 countries, including Costa Rica, Nicaragua, and Uruguay, met 10 per cent of their demand from wind. In 2018, solar PV accounted for 12.1 per cent of total generation in Honduras. Significant shares can also be observed in Italy and Greece (both about 8.2 per cent), and by late 2018 one in five Australian households generated at least some of their electricity with solar energy (Ren21 2019).

Bloomberg New Energy Finance (BNEF) has published its analysis for investment in global clean energy which shows that 2017 was the second highest ever, with US$333.5 billion, despite falling technology costs (Louw 2018). Globally, the solar sector in China dominated, with a total of US$132.6 billion of investments – leading to over 50 GW of additional solar capacity. As the figure 1

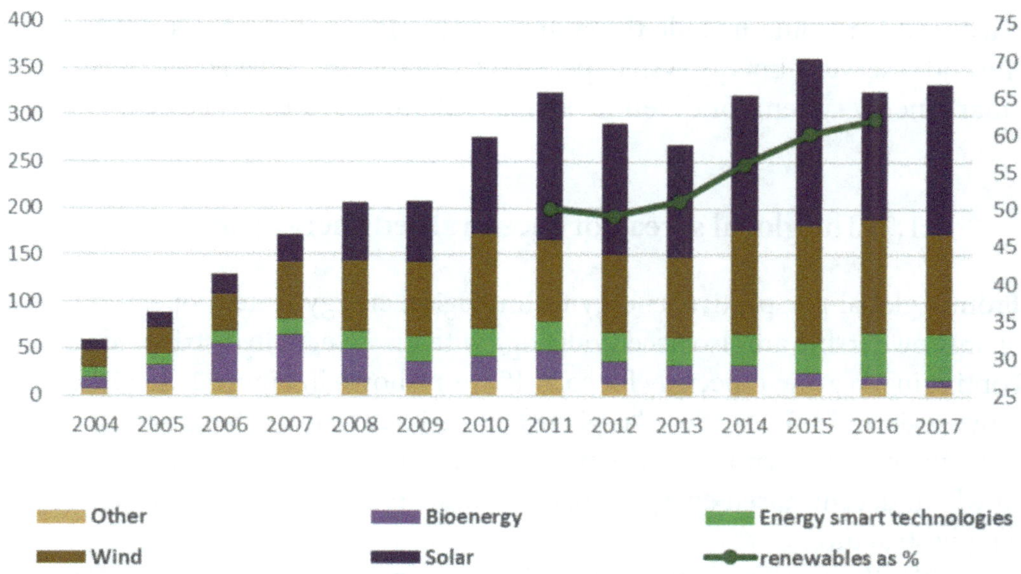

Figure 1: Global investment in clean energy by sector (US$ billion).
Source: BNEF (2018), UNEP 2010–2017 Status Reports.

shows, growth in solar investment has continued at pace over the last decade. The graphic also indicates the extent to which the deployment of renewable energy, including hydropower, has come to dominate total new capacity in the power sector, moving from around 50 per cent at the turn of the century to, in 2016, comprising of 62 per cent.

In regional terms Asia, largely China, dominates the global landscape, with Europe, once the world leader, continuing to decrease the level of investments. In 2017, European investment totalled US$57.4 billion, down from US$137.8 billion in 2011. In the United States, investment in clean energy grew marginally to US$56.9 billion in 2017, with a peak in 2011, similar to Europe, in the United States at US$62.3 billion. Mexico and Australia saw 2017 investment levels at an all-time high of US$6.2 billion and US$9 billion, respectively.

In terms of installed capacity and output, the European Union still is a global leader in renewable energy. In 2017 across the bloc, renewables, including hydropower (9 per cent), renewables provided 33 per cent of electricity, more than any other source (Sandbag 2019).

While renewable energy deployment has been initiated by national policies and measures, with 179 countries having renewable energy targets on the national or state or provincial level (Ren21 2019), this is expected to accelerate, as equipment and production costs for small-scale renewables continue to fall and reach grid parity[1] in many regions of the world. However, this is only the first step. The next step is when renewables are able to achieve 'energy system

[1] Grid parity (or socket parity) occurs when an alternative energy source can generate power at a levelised cost of electricity (LCOE) that is less than or equal to the price of purchasing power from the electricity grid.

parity', which would include the system integration costs (the costs of balancing and reserves). Energy system parity is likely to be achieved once integrated, smart energy systems or decentral storage solutions come into place.

1.2 The global spread of decentralised energy generation

From a global perspective, energy technologies, energy system operation, and energy ownership are also decentralising,[2] with investment in distributed energy continuing to grow, especially for solar PV, as is shown in Figure 2. In 2017, both large-scale and small-scale solar picked up again close to 2015 figures.

As the country reports in Chapter 2 demonstrate, there are some countries in which decentralised energy is playing an increasingly important role in the supply structure.

In 2017 the worldwide investment in solar projects of less than 1 MW was US$49.4 billion, installing 29 GW. China rapidly increased its investment five-fold in 2017, totalling US$19.6 billion of investment in small-scale projects, almost 40% of the global total. While the global investment in small-scale renewables is much less than the peak in 2011, of US$76.2 billion, as the cost of solar has fallen 57% over the same period, the annual installed capacity is the largest yet (Frankfurt School-UNEP 2018).

For a couple of years, Japan dominated the country ranking of investments in decentralised renewable energies, with a total of US$31.7 billion in 2015. The rise of renewable deployment in Japan was, in part, a response to the accident at the Fukushima nuclear power plant in 2011 and the subsequent temporary closure of all nuclear power plants. Since then, much to the discomfort of the national government, restart of the reactors has been extremely slow. At the end of the year 2018 only nine plants were in operation, down from 54 prior to the accident.

Australia has seen rapid growth in the deployment of solar, especially on the household level. Despite cuts in government support, deployment of PV has continued, because its decreasing costs turn them into economically attractive alternatives to paying the retail price of electricity. By the end of 2017,

[2] By decentralising we mean: technologies themselves are in smaller capacity units, and their geographic distribution is wider. The system is moving from a one-way, top-down, supply-orientated operation of a few, large conventional fossil power plants to a system operation in a bi-directional way, demand focused, with multiple, varied power generating units of all sizes. Whereas ownership used to be state-owned monopolies, or large utilities, increasingly there are new entrants with non-traditional business models which provide particular services, for example suppliers that only sell renewable energy; former municipal utilities that diversify their services; independent platform providers which establish local energy markets; intermediaries who manage demand-side response.

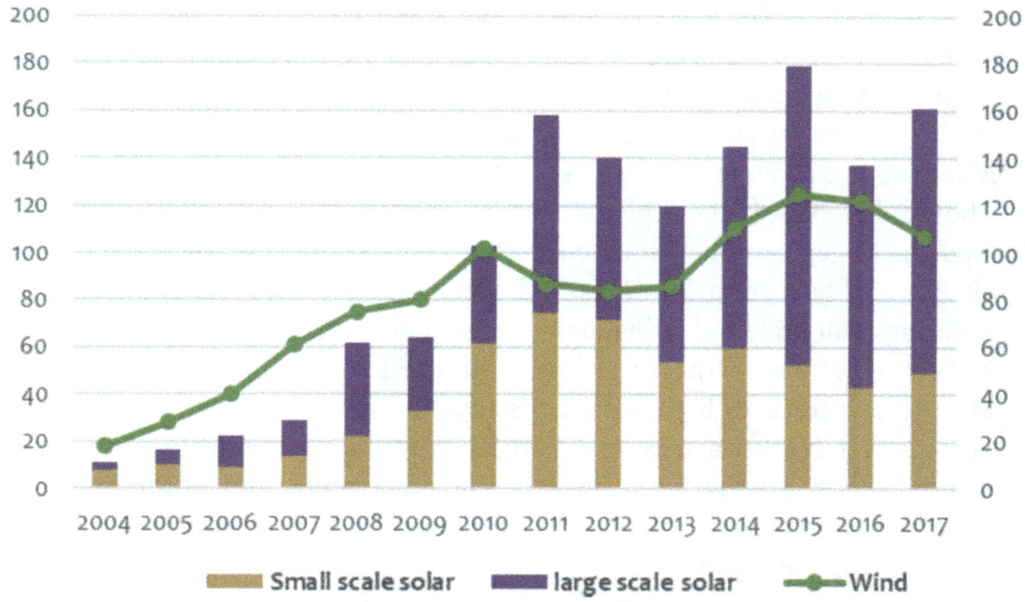

Figure 2: The global deployment of solar and wind power (GW).
Source: BNEF 2018.

cumulative installed capacity for solar PV systems in Australia stood at 6401 MW with close to 1.8 million installations, an increase from 5463 MW and 1.64 million installations the previous year (AEC 2018). By the start of 2018, over 30 per cent of homes in Queensland and South Australia had solar panels (Australian PV Institute 2018).

Across Europe around two thirds of solar systems are located on the rooftops, be they residential, commercial, or industrial. In Germany, there are approximately 1.5 million solar PV systems (GTAI 2018) with a total installed capacity of nearly 43 GW by the end of 2017, but only a small amount of utility-scale solar units. By contrast, utility-scale solar accounts for 20 per cent of the approximately 20 GW PV capacity within the Italian system – there are still over 700,000 separate solar installations (Gianni 2017).

In the United States in 2017, around 28.5 GW of electricity generating infrastructure was deployed – 25 GW utility scale and about 3.5 GW of distributed (that is, smaller than 1 MW) solar power – of the total, wind and solar were 55.4 per cent. However, when looking at net additions with the closure of 11.8 GW of utility-scale fossil fuel plant retirements, the net new volume of US generation was 16.7 GW of generating capacity, with 94.7 per cent of that coming from renewables (Weaver 2018). In comparison, India's rooftop solar accounts for 9 per cent of the country's solar capacity. In Japan, around 11.8 per cent of new solar additions are on rooftops (REN21 2018).

China has experienced a massive increase in deployment of decentralised renewable technologies. Over the last few years, China has shown that without the engagement of customers and the public, renewable energy, in particular solar and wind, can be deployed at scale too. These deployment rates, 53 GW in 2017 alone, have had a profound impact on global technological manufacturing costs. There has also been a shift towards distributed capacity, with about 19.4 GW of capacity added in 2017, up from 4.2 GW in 2016, including a threefold increase in rooftop solar to 2 GW (REN21 2018).

The rapid increase in the deployment of renewables has most often been driven by specific targets or policy interventions. In the case of Germany and Italy, the availability of feed-in-tariffs (FiTs) led a boom in PV deployment, including significant small-scale and individually or community owned. In some countries, the effect of price guarantees was underestimated. Programs were exploited in a short time leading to high overall costs for the support schemes and started to affect other market actors, traditional generating companies, and the grid operators. Consequently, fiscal support schemes – reductions in the FiTs and more recently changes in rules about grid access – have slowed done, and in some cases completely stopped.

By contrast, the developing world is leapfrogging into a decentralised energy supply infrastructure. In developing countries, micro-grids and solar-storage kits for individual households co-exist at the periphery of the central grid and may in future substitute the rollout of the public transmission network, comparable to the phenomenon of leapfrogging from no telephone service to handheld devices without passing the stage of line-based telephony.

1.3 Decentralised storage gaining importance

The greater deployment of renewables, particularly those with weather-dependent, variable production, is increasing the need for grid flexibility and reducing the need for traditional base-load generators. A key technology to increase flexibility is storage technology. An assessment by the US Department of Energy suggests that storage will increase the possibility of economic deployment of variable renewables from 16 per cent to 55 per cent (NREL 2016).

Storage technologies will also enable the greater use of electricity in other sectors, such as heat and transport. Recognising these cross-sector benefits has resulted in increased efforts in research and development, leading to greater deployment and creating a virtuous circle of falling, higher technical potentials, and further deployment.

Advances in storage technologies are especially important for electric vehicles, as they face the trade-off between weight of the batteries and restrictions in the range, which may lead to so-called 'range anxiety' of drivers. Nonetheless, the race to electrify the transport sector is speeding up. Significantly, Volvo, the Chinese-owned Swedish car manufacturer, announced in June 2017 that all its cars built after 2019 would be hybrid or purely electric, the first major automotive firm to do so (Vaughan 2017). Bloomberg New Energy Finance have revised their forecasts and have suggested that, by 2040, 57 per cent of all new car sales will be electric and that electric vehicles (EVs) are expected to be at parity with internal combustion vehicles by the mid-2020s in most markets (BNEF 2019). However, some countries are likely to move much quicker than the BNEF global average, with France and the United Kingdom announcing that they will ban the sale of petrol and diesel cars by 2040 (Chrisafis & Vaughan 2017). The rollout of electric vehicles will have profound impact on the power sector, through increased and flexible demand, cheaper electric storage technology, and the cross-over between actors in the utility and car manufacturing markets. Directly competing with premium EV manufacturer Tesla, traditional car producers such as BMW, Honda, and Nissan have already started selling household-level storage units, both to capitalise on their existing battery research, but also as a potential use for second-life batteries.[3]

This combined potential use has resulted in overall cost reduction in lithium batteries that are in line with those seen in the wind and solar PV sectors. The cost of the latest electric vehicle by car manufacturer Tesla, which entered production in 2017, has costs of US$190 per kWh (Voelcker 2016), BNEF expect that costs will continue to decline reaching as low as US$70 per kWh by 2030 (BNEF 2018). In the power sector, the costs for consumer level (Lambert 2016) or grid level storage are also falling fast, helping to accelerate their rate of deployment.

Progress in developing and commercialising new storage technologies, in particular solid-state batteries with a higher energy intensity than lithium-ion batteries and less use of scarce raw materials, is likely to accelerate the usage of batteries not only in automobiles, but also in applications around the smart home (Forschungszentrum Juelich 2018).

1.4 Decoupling growth and energy intensity via renewables and energy efficiency

Renewable deployment and storage technologies must go hand in hand with energy efficiency, if the system is to meet overall objectives of decarbonisation. The developments in energy demand vary hugely across the world's economies. The most striking feature in recent decades has been the increase in consumption in China and India, as can be seen in Figure 3. Although China's consumption is now three times larger than it was at the turn of the century, the overall growth rates seem to decrease, which is the case for energy, as seen in the graphic, but also for electricity. While the growth in India has been slower, there is currently no tapering off, with an expectation that by 2035 it will exceed that of China's (BP 2017).

[3] Once batteries have degraded to some 80 per cent of their capacity, they may longer be suitable for vehicle usage, but may still be suitable to stationary storage where size and weight are no longer critical factors.

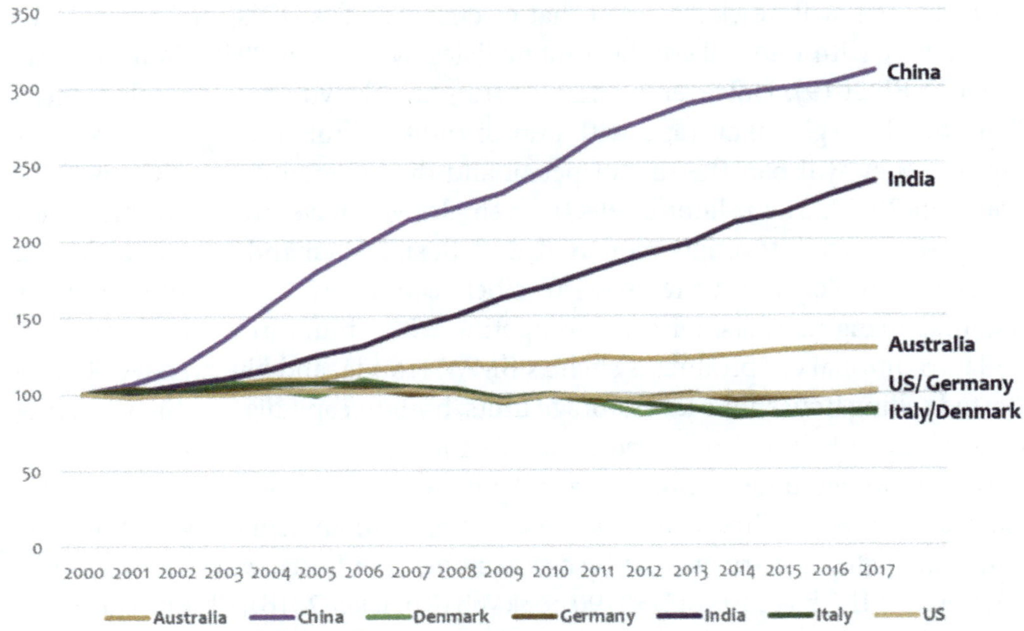

Figure 3: Relative increase in energy consumption in selected countries.
Source: BP Statistic Review (2018).

The energy intensity of the global economy is decreasing due to technological progress and systemic changes. Despite rising GDPs in many OECD countries, demand for energy and electricity are stable or falling, also because of structural changes in their economies – with less reliance on energy-intensive industries and a shift to services and digital production, as is shown in Figure 4.

In emerging economies, increase in energy demand has slowed significantly, mainly due to improved efficiencies, a reduced rate of infrastructure construction, for example, use of cement, and to some extent the increasing role of the service sector.

1.5 Value creation with decentralised renewable energy generation

Many countries still rely on a fully regulated electricity supply industry, often with vertically integrated utilities and a single-buyer model. For these countries, one driver for a stronger push towards decentralised supply structures may be motivated by over-arching policy objectives, namely local value creation and employment.

IRENA estimate that in 2016 renewable energy employed 9.8 million people, of which 3.1 million were in photovoltaics sectors and 1.2 million in wind power. Globally, China accounts for 3.6 million of the global jobs, of which solar PV accounted for 2.0 million. Within the sector, 1.3 million were for the

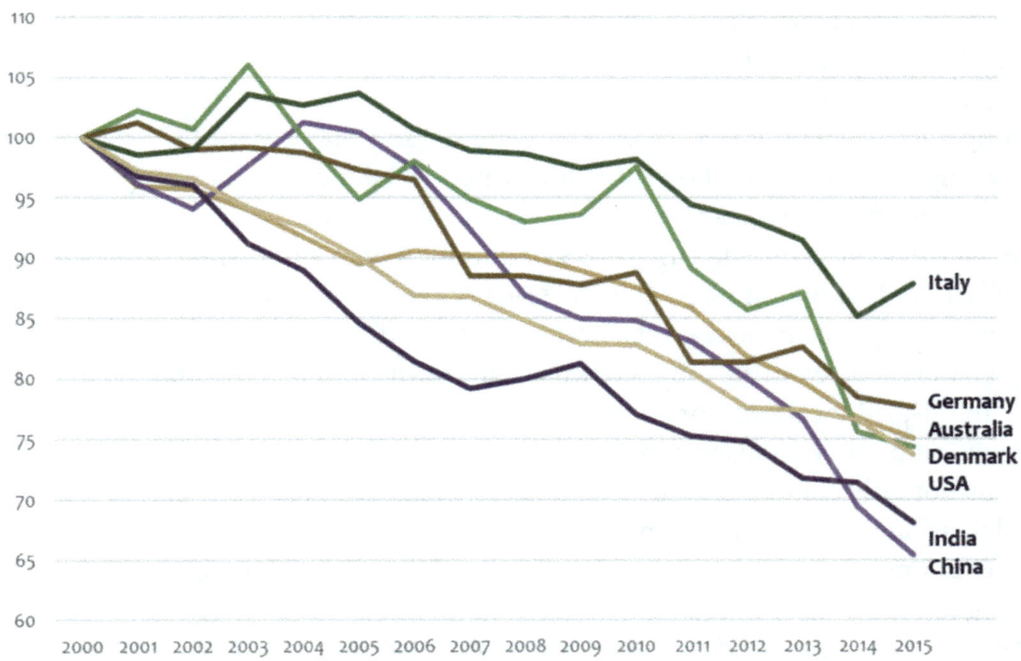

Figure 4: Relative change in energy intensities 2000–2015.
Source: World Bank (2018).

manufacturing of the PV panels, with 635,000 for construction and installation and 26,000 for operation and maintenance (IRENA 2017a). In Bangladesh, solar photovoltaics accounted for 140,000 jobs (ibid.). On a global basis across the value chain of a medium sized solar installation (50 MW), 22 per cent of the jobs are related to construction, while 17 per cent are in the installation, and 56 per cent in the operation and maintenance (IRENA 2017b).

An analysis of the US Department of Energy reveals that employment in the solar and wind industries totalled 374,000 and 102,000 individuals, respectively, out of a total workforce of 1.9 million in electric power generation and fuels technologies. Compared to 2015, employment in solar and wind industries increased by 25 per cent and 32 per cent, respectively, in 2016 (US Department of Energy 2017). If these figures are combined with the energy output by technology, they translate into 7724 MW hours per worker in the case of coal, 3812 MW hours per worker for natural gas, and 98 MW hours per worker for solar (Perry 2017). As Perry states, 'to produce the same amount of electric power as just one coal worker would require two natural gas workers and an amazingly high 79 solar workers' (ibid.) Perry interprets these comparative figures as an indicator for the lack of productivity in generating electricity from renewable resources, but if a government's emphasis is on communal value creation, the new technologies offer unprecedented opportunities for local employment.

For the year 2012, the independent research institute IÖW investigated the impact of renewable energies on local value creation in Germany. With 36 per cent, the manufacturing of components has the highest overall share in value creation, but this activity does only marginally take place on the municipal or regional level. However, the other elements of value creation, including planning and installation (13 per cent), operation and maintenance (23 per cent), trading (6 per cent), and revenues for owners (22 per cent), are strongly tied to the location of the installations: 'The direct value added by renewable energies in Germany in 2012 adds up to 16.9 billion EUR with a municipal value added of around 11.1 billion EUR. Therefore 66 per cent of the total value added in the Federal Republic benefit local communities. In addition to that, nearly 380,000 jobs were created in 2012 by renewable energy in Germany' (Aretz et al. 2013).

In off-grid locations, for example in Sub-Saharan Africa, the effects of value creation by decentralised supply may be even more pronounced, because they may not only contain the direct financial benefits for local communities, technicians, and owners of the installations, but may also include indirect effects. For example, they may positively affect quality of life in impoverished rural neighbourhoods, thus reducing urban migration and brain drain. In Chapter 3, the business models of start-ups Mobisol, Solarkiosk, and SolShare are presented, which target off-grid communities and enhance the local economy.

1.6 Digitalisation as enabler of the smart grid and new business models

In all facets of our lives we have entered the digital age. Not only communication, social interaction, and entertainment, but also shopping via the internet, the smart home with assisted living, e-government, or individual mobility with autonomous and connected cars. New information technologies and systems are revolutionising the energy sector, too, through the generation of individual data and the ability to process and analyse it, to the opportunities for machine learning, improving energy performance, and the increasing use of distributed ledger technologies, in particular Blockchain.

While the electricity supply industry has been slow to become aware of the opportunities and threats that these data management technologies can bring, the degree of interest and speed at which pilots are being undertaken indicate that rapid change is likely. Predictive maintenance of devices, such as turbines in thermal power plants, and more precise forecasts of consumption patterns have already become reality. The next step for data management will be to integrate micro-producers of electricity – so-called prosumers – into the balancing of the distribution network.

The rollout of smart meters and the installation of smart devices, such as sensors, in transmission and distribution networks is rapidly increasing the volume of information and the ability to process this. On average, a smart meter recording every 15 minutes transmits 400 MB of data each year. This information serves as an enabler of new markets, allowing businesses to collect, anonymise, and analyse it, to potentially increase efficiency, better match supply with demand, and also – vice versa – enable demand to increasingly match variable supply.

However, smart meters and the linking of a wide range of electronic devices raises security concerns, specifically over privacy implications, despite suggestions that these fears maybe over-stated (Wang & Lewandowski 2016; Burger, Trbovich & Weinmann 2018).

Machine learning, whereby computers can improve their decision-making capabilities with minimal human intervention, has been part of the development of information technologies since the 1950s. However, the success of its usage and its potential have only been recently widely recognised in the energy sector. For example, machine learning is envisaged to improve the efficiency of generation, both conventional and renewable (Murgia & Thomas 2017). By machine learning with their Deep Mind computer power, Google estimates that it saves 40 per cent of consumption by optimising the efficiency of operation and predicting future data and energy usage. Information technology has become a major consumer of energy, but Google claims that because of super-efficient servers and rapid improvements in computer power they have increased the level of computer power they can produce per unit of energy consumed by 3.5 times over the last five years (Deepmind 2016).

Blockchain may be another source of digital disruption of the energy sector. It is a distributed database of data records that links transactions to each other, thereby providing transparency. Blocks are verified by a distributed network of computers. Transactions are encrypted. This cuts out the middlemen, allowing not only payment transactions but also smart contracts, the technology might provide a basis to embed prosumers into the energy system and deal with the resulting increasing complexity by reducing process costs and enabling platforms for smart contracts beyond a single energy provider (Burger et al. 2016; PwC 2016).

The above-mentioned key trends of competitiveness of renewables, the rising role of decentralised supply and storage, decoupling growth and energy intensity, enhancing local value creation, and digitalisation as enabler of smart grids and new business models describe the uncertainty policy makers and company leaders are facing. By looking at cases of country transformations and business models beyond subsidies, the next two chapters build a basis and guide through this uncertainty, including the necessary changes for governance, as introduced in the next chapter.

1.7 References

AEC. (2018), *Solar report, January 2018, Australian Energy Council.* Retrieved September 18, 2018, from https://www.energycouncil.com.au/media/11188/australian-energy-council-solar-report_-january-2018.pdf

Aretz, A., Heinbach, K., Hirschl, B. & Schröder, A. (2013), *Wertschöpfungs- und Beschäftigungseffekte durch den Ausbau Erneuerbarer Energien*. Studie im Auftrag von Greenpeace Deutschland, Hamburg. Berlin, Institut für ökologische Wirtschaftsforschung (IÖW)

Australian PV Institute. (2018), *Mapping Austrailian photovoltaic installation*. Retrieved September 18, 2018, from http://pv-map.apvi.org.au/historical#4/-26.67/134.12

Bellini, E. (2017), Chile's auction concludes with average price of $32.5/MWh, *PV magazine International*, 3 November 2017, see https://www.pv-magazine.com/2017/11/03/chiles-auction-concludes-with-average-price-of-32-5mwh/

Bhaskar, U. (2017), India's wind power tariff falls to a record low of Rs2.64 per unit, *Livemint*, 6 October, https://www.livemint.com/Industry/sMC62YoWv4LybaPZnKivMM/Indias-wind-power-tariff-falls-to-a-record-low-of-Rs264-pe.html

BNEF. (2019), Electric vehicle outlook, 2019, Bloomberg new Energy Finance. Retrieved from https://about.bnef.com/electric-vehicle-outlook/

BNEF. (2018), *Electric vehicle outlook, 2018, Bloomberg new Energy Finance*. Retrieved February 1, 2019, from https://about.bnef.com/electric-vehicle-outlook/

BP. (2017), *BP energy outlook 2017*. London, United Kingdom.

BP. (2018), *BP statistical review of world energy 2018*. London, United Kingdom.

Bundesnetzagentur. (2017), Results of first auction for onshore wind installations, 19 May 2017, https://www.bundesnetzagentur.de/SharedDocs/Pressemitteilungen/EN/2017/19052017_onshore.html?nn=404422

Burger, C., Kuhlmann, A., Richard, P. & Weinmann, J. (2016), Blockchain in the energy transition. A survey among decision-makers in the German energy industry. Berlin: German Energy Agency dena/ESMT. Retrieved from https://shop.dena.de/sortiment/detail/produkt/studie-blockchain-in-der-energiewende/

Burger, C., Trbovich, A. & Weinmann, J. (2018), Vulnerabilities in smart meter infrastructure – can blockchain provide a solution? Results from a panel discussion at EventHorizon2017. Berlin: German Energy Agency dena/ESMT. Retrieved from https://press.esmt.org/all-press-releases/blockchain-can-improve-data-security-energy-infrastructure

Chrisafis, A. & Vaughan, A. (2017, July 6), France to ban sales of petrol and diesel cars by 2040. *The Guardian*. Retrieved February 1, 2019, from https://www.theguardian.com/business/2017/jul/06/france-ban-petrol-diesel-cars-2040-emmanuel-macron-volvo

Deepmind. (2016), *Deepmind AI reduces Google data centre cooling bill by 40 per cent*. Retrieved August 3, 2017, from https://deepmind.com/blog/deepmind-ai-reduces-google-data-centre-cooling-bill-40/

Dudley, D. (2019), Race heats up for title of cheapest solar energy in the world, *Forbes*, 17 October 2019, https://www.forbes.com/sites/dominicdudley/2019/10/17/cheapest-solar-energy-in-the-world/#2f35ffd94772

Forschungszentrum Juelich. (2018), Toward fast-charging solid-state batteries. *ScienceDaily*. Retrieved August 20, 2018, from www.sciencedaily.com/releases/2018/08/180820094448.htm

Frankfurt School-UNEP Centre/BNEF. (2018), *Global trends in renewable energy investment 2018*. Retrieved February 1, 2019, from http://www.fs-unep-centre.org

Gianni, M. (2017, September), *PV market, business and price developments in Italy, Gestore Servizi Energetici*. Presented at the EU PCSEC 2017. Retrieved September 18, 2017, from https://www.gse.it/documenti_site/Documenti%20GSE/Studi%20e%20scenari/PV%20market,%20business%20and%20price%20developments%20in%20Italy.PDF

GTAI. (2018), *Photovoltaics, ergmany trade and investment*. Retrieved September 18, 2018, from (http://www.gtai.de/GTAI/Navigation/EN/Invest/Industries/Energy/photovoltaic.html

Henbest, S. (2017), *Global key message from New Energy Outlook, Bloomberg New Energy Finance*. Retrieved August 2, 2017, from https://data.bloomberglp.com/bnef/sites/14/2017/06/NEO-2017_CSIS_2017-06-20.pdf

Hill, J. (2018), Germany Awards 900 Megawatts In Wind & Solar Tenders In Year's First Auctions, *CleanTechnica*, 23 February 2018, see https://cleantechnica.com/2018/02/23/germany-awards-900-mw-wind-solar-tenders-years-first-auctions/.

Hill, J.S. (2019), Saudi Arabia wind farm claims world record low energy cost, *RenewEconomy*, 13 August 2019, see https://reneweconomy.com.au/saudi-arabia-wind-farm-claims-world-record-low-energy-cost-99966/

Hill, J.S. (2019), UK offshore wind prices reach new record low in latest CfD auction, *CleanTechnica*, 23 September 2019 https://cleantechnica.com/2019/09/23/uk-offshore-wind-prices-reach-new-record-low-in-latest-cfd-auction/

IRENA. (2017a), *Renewable energy and jobs, annual review 2017, International Renewable Energy Agency*. Retrieved from http://www.irena.org/publications/2017/May/Renewable-Energy-and-Jobs--Annual-Review-2017

IRENA. (2017b), *Renewable energy benefits, leveraging local capacity for solar PV, International Renewable Energy Agency*. Retrieved from http://www.irena.org/publications/2017/Jun/Renewable-Energy-Benefits-Leveraging-Local-Capacity-for-Solar-PV

IRENA. (2019), Renewable power generation costs in 2018, *International Renewable Energy Agency*. Retrieved from https://www.irena.org/publications/2019/May/Renewable-power-generation-costs-in-2018

Klessmann, C. & Tiedemann, S. (2017), *Germany's first renewables auctions are a success, but new rules are upsetting the market*. Retrieved from http://energypost.eu/germanys-first-renewables-auctions-are-a-success-but-new-rules-are-upsetting-the-market/

Lambert, F. (2016), *Tesla quietly reduced the price of the Powerpack by 5 per cent and its commercial investor by 19 per cent, electrek*. Retrieved from https://

electrek.co/2016/09/09/tesla-quietly-reduced-the-price-of-the-powerpack-by-5-and-its-commercial-inverter-by-19/

Lillian, B. (2019), DOE report confirms wind energy costs at all-time lows, *North America Windpower*, 15 August 2019, see https://nawindpower.com/doe-report-confirms-wind-energy-costs-at-all-time-lows

Louw, A. (2018), *Clean energy investment trends, 2017.* Retrieved January 16, 2018, from https://data.bloomberglp.com/bnef/sites/14/2018/01/BNEF-Clean-Energy-Investment-Investment-Trends-2017.pdf

Murgia, M. & Thomas, N. (2017, March 12), DeepMind and National Grid in AI talks to balance energy supply. *Financial Times.* Retreived June 22, 2017, from https://www.ft.com/content/27c8aea0-06a9-11e7-97d1-5e720a26771b?mhq5j=e3

NREL. (2016), *Energy storage, possibilities for expanding electric grid flexibility, National Renewable Energy Laboratory.* Retrieved from http://www.nrel.gov/docs/fy16osti/64764.pdf

Perry, M.J. (2017), *Inconvenient energy fact: It takes 79 solar workers to produce same amount of electric power as one coal worker. AEIdeas, American Enterprise Institute.* Retrieved from http://www.aei.org/publication/inconvenient-energy-fact-it-takes-79-solar-workers-to-produce-same-amount-of-electric-power-as-one-coal-worker/

Pratheeksha. (2019), How Low Did It Go: 5 Lowest Solar, Tariffs Quoted in 2018, *Mercom Communications India*, 8th January 2019, https://mercomindia.com/lowest-solar-tariffs-quoted-2018/

PwC. (2016), Blockchain – an opportunity for energy producers and consumers? (A study conducted by PwC on behalf of Verbraucherzentrale NRW, Düsseldorf, PwC global power and utilities)

Ren21. (2019), Renewables 2019 global status report, renewable energy policy network for the 21st Century. Retrieved from https://www.ren21.net/wp-content/uploads/2019/05/gsr_2019_full_report_en.pdf

Sandbag. (2019), The European power sector in 2018, state of affairs and review of current development analysis. Agora and Sandbag, January 2019. Retrieved from https://sandbag.org.uk/project/power-2018/

Saurabh. (2017), New low for India wind energy tariff In 500 Megawatt auction, *Cleantechnica* December 25th, 2017, see https://cleantechnica.com/2017/12/25/new-low-india-wind-energy-tariff-500-megawatt-auction/

Schneider, M. & Froggatt, A. (2019), World nuclear industry status report, September 2019. Retrieved from https://www.worldnuclearreport.org/

Spatuzza, A. (2018), Brazil contracts 114MW of wind in tender at record low prices, *Recharge*, 4 April 2018, https://www.rechargenews.com/wind/brazil-contracts-114mw-of-wind-in-tender-at-record-low-prices/2-1-308198

US EIA. (2017), *More than half of small-scale photovoltaic generation comes from residential rooftops, 1 June 2017, US Energy Information Administration.* Retrieved from https://www.eia.gov/todayinenergy/detail.php?id=31452

Vaughan, A. (2017), All Volvo cars to be electric or hybrid from 2019. *The Guardian*. Retrieved February 1, 2019, from https://www.theguardian.com/business/2017/jul/05/volvo-cars-electric-hybrid-2019

Voelcker, J. (2016), *Electric-car battery costs: Tesla $190 per kwh for pack, GM $145 for cells, Green Car Reports, 28 April 2016*. Retrieved December 1, 2016, from http://www.greencarreports.com/news/1103667_electric-car-battery-costs-tesla-190-per-kwh-for-pack-gm-145-for-cells

Wang, Q. & Lewandowski, S. (2016, July), *Are smart meters being used smartly? A case study of residential electricity customers in Vermont*. Selected paper prepared for presentation at the 2016 Agricultural and Applied Economics Association Annual Meeting, Boston, Massachusetts, July 31 August 2 2016. Retrieved from https://ageconsearch.umn.edu/record/236144/files/Are%20Smart%20Meters%20Being%20Used%20Smartly%20-%20A%20Case%20Study%20of%20Residential%20Electricity%20Customers%20in%20Vermont.pdf

Weaver, J. (2018), *More than 94 per cent of net new electricity capacity in the USA from renewables in 2017 – emissions down 1 per cent, Electrek, 12 January 2018*. Retrieved June 7, 2018, from https://electrek.co/2018/01/12/94-percent-new-electricity-capacity-usa-from-renewables/

Weaver, J. (2019), Los Angeles seeks record setting solar power price under 2c/kWh, *PV magazine*, 28 June 2019. https://pv-magazine-usa.com/2019/06/28/los-angeles-seeks-record-setting-solar-power-price-under-2%C2%A2-kwh/

Wood Mackenzie. (2019), Portugal's world record-breaking solar PV auction, 12 August 2019, https://www.woodmac.com/press-releases/portugals-world-record-breaking-solar-pv-auction/

CHAPTER 2

Regulatory and policy incentives – how to establish governance for decentralised energy systems?

Antony Froggatt and Catherine Mitchell

2.1 The role of regulation and governance

2.1.1 Regulation as accelerator or decelerator of the energy transformation

There are multiple ways to meet the 2°C reduction target of the Paris Agreement and to reach a 'deep decarbonisation' of our economies, including a reduction of primary energy consumption, the use of low-emission generation technologies such as nuclear power, or the use of carbon capture and storage. However, the most likely and cost-effective path of decarbonisation is that of renewables supplying the majority of electricity, if not the entirety (GEA 2012; IPCC 2015; Greenpeace 2015) alongside significant energy efficiency measures, whether minimising energy use in buildings through retrofit programmes or via the markets, ensuring the demand side is as valuable as the supply side. That will require not only new policies and significant changes in incentive schemes for generators and the associated grid infrastructure, but also a more

How to cite this book chapter:
Froggatt, A. and Mitchell, C. 2020. Regulatory and policy incentives – how to establish governance for decentralised energy systems? In: Burger, C., Froggatt, A., Mitchell, C. and Weinmann, J. (eds.) *Decentralised Energy — a Global Game Changer.* Pp. 21–24. London: Ubiquity Press. DOI: https://doi.org/10.5334/bcf.b.

encompassing transformation of governance mechanisms – policies, institutions, market design and network rules, and the 'politics' behind them.

The starting point of today's regulatory framework was centralised systems. Historically, only one, or a few, entities owned and operated the grid and its associated infrastructure. Customers were passive receivers of a public infrastructure service. Many countries have not yet undertaken any reforms in the power supply industry at all, and vertically integrated, state-owned utilities are in charge of all operations along the value chain of the power sector. In other countries, reforms have been initiated, such as in South Korea, where competition is in principle allowed, but a single state owned company still dominates the market.

As described above, many countries are experiencing a phase of rapid decentralisation, however. This means new roles for, and new relationships between, stakeholders – whether resource providers, buyers, transmission operators, and so on. The idea of 'grids' will have to alter. They will be extended, interconnected, and more dynamic to operate flexibly, thereby incorporating variable renewables most cost effectively. Flexibility may become a key system function affecting large groups of dispersed individual suppliers.

In competitive electricity markets, such as in Western and Northern Europe, different customer segments emerge. Some individual consumers, as well as groups of consumers such as consumer co-operatives, may become producers and investors, and get involved in managing grids, whereas other consumers may choose to continue in their traditional role as customers and recipients of energy services. Some residential consumers may not even have an interest in switching their supplier, as their energy bill is just a small share of their income. For example, in Denmark – despite being a country where customers are invested in community projects – households have not tended to change electricity suppliers.

In situations where both individuals and groups of customers choose to become active agents, commercial opportunities for new entrants to provide services also occur. However, the energy system and its operations, the coordination and integration of smaller-scale services, and dispersed ownership, are also becoming more complex. Not only does it require greater flexibility for efficient operation, but it also requires greater data transparency and processing power of that data.

2.1.2 An assessment of governance practices in key transformation countries

Decentralising the energy system and establishing local markets, including local balancing markets, provide commercial opportunities for customers or producers, thereby adding a new dimension to the energy system – distributed, decarbonised, digitalised, and potentially democratised. Therefore, the role of regulation will change. It needs to be able to assign value to new services, to create a means of coordination between networks, markets, and new platforms in ways that are cost effective to customers, but also nimble and adaptive enough to enable, rather than undermine or block, innovation, new business models, and customer wishes.

The business models of these developing energy systems will differ depending on geography and sociocultural context, the legacies of the previous systemic configurations, and the size and pattern of demand. However, the way the business models develop in these different types of energy systems is also determined by where value – or payments – can be accessed within the energy value chain; and this in turn depends on the extent to which governance enables, constrains, or channels energy system innovation. Governance is at the centre of energy system transformation including the rate at which it is able to decarbonise.

The deployment and the development of decentralised energy differs across countries and regions, depending on the policy situation in which they develop. Within each country there are often distinct periods of renewables deployment, driven by different priorities in the policy regimes:

- Some countries, such as Denmark and Germany, which both have a long history of renewable energy deployment, have put renewables at the centre of their energy and electricity policy. The high level of deployment subsequently implies that an efficient integration of current and future volumes of renewables requires an increase in the flexibility of their energy systems.
- Emerging economies – such as China and India, characterised by rapidly growing power demand – have become global leaders in renewable technologies, including both manufacturing and deployment. These unprecedented high annual renewable installation rates may make small differences to their countries' overall power mixes, given the size of the existing supply portfolio, but the scale of these investments have indeed affected technology adoption internationally by driving down the price.
- There are regions in which renewable deployment is already affecting grid operation, such as South Australia or California.
- In other countries, where renewables play a less significant role, institutions have been slower to reform, so that the governance structure itself is becoming a limiting factor in the efficient deployment of decentralised renewables. For example, the Italian market is characterised by a remarkable increase in decentralised renewable capacity installed, despite the dominance of former incumbent utility Enel and the hesitation of regulatory bodies to introduce a coherent regulatory regime that embraces the next wave of the transition towards a fully decentralised system.

This chapter of the book looks at the current deployment practises in these countries, to assess the effectiveness of the policies and the impacts that higher contributions of renewables and decentralisation are having on system operation. The countries and regions were selected based on their historic, current, or expected importance to the global deployment rate of renewable energy and the contribution that renewables make to the overall electricity supply, and the system consequences. Some countries, for example the United States, which has a federal political model, have multiple, differing policies in different devolved states, which can illuminate the value of different governance mechanisms. The US chapter has been written as a comparative chapter in order to reflect this diversity. These country chapters can be allocated into two main categories:

- Countries or states with a *high* share of new renewable energies, but *low or stagnant* growth of per-capita primary energy consumption:
  - Australia
  - California and New York State
  - Denmark
  - Germany
  - Italy
- Countries or states with a *low* share of renewable energies and currently *low levels* of per-capita primary energy consumption, but on a fast economic development path:
  - China
  - India

The following sections of this chapter depict the diverging options governments have chosen to promote decentralised renewable energy supply. All contributions are independent yet interdependent narratives of the global momentum of the transformation of the energy supply industry. All of them emphasise the importance of regulation in the systemic changes, but they also show how the spectrum of trajectories – and their outcomes – are embedded in a path-dependent sociocultural and economic ecosystem that differs fundamentally across nations and continents.

2.1.3 References

GEA. (2012), Global energy assessment towards a sustainable future, GEA and International Institute for Applied Systems Analysis. Cambridge: Cambridge University Press.

Greenpeace. (2015), Energy [r]evolution – A sustainable world energy outlook, 2015, 100 percent renewable energy for all, 5th Edition.

IPCC. (2015), Renewable energy sources and climate change mitigation (Special report of the Intergovernmental Panel on Climate Change. Bonn: Intergovernmental Panel on Climate Change).

2.2 Australia: from central electricity to solar/storage systems

Helen Poulter

2.2.1 *Introduction*

Australia has one of the best solar resources in the world and also some of the highest electricity prices. It also has the highest per capita of domestic solar PV installations worldwide, at almost 25 per cent of households. In September of 2016, South Australia experienced severe storms which led to a state-wide blackout and prompted a review of the governance of the National Electricity Market (NEM). A heatwave in February 2017 caused the NEM to cut power to 90,000 of its domestic customers. This load shedding event prompted a further peak in the installation of behind-the-meter distributed generation and enquiries for domestic storage.

This chapter will give an overview of the NEM in Australia, the renewable energy resources, the current situation regarding distributed energy, and the future plans for the governance of the NEM.

2.2.2 *The Australian electricity framework*

Australia's electricity networks are separated into three regional markets and one state owned system. The NEM is the largest and operates in Eastern Australia, covering five interconnected state-based networks and is the focus of this chapter. The three other smaller systems operating in the western and northern states (The Wholesale Electricity Market (WEM), the Northern Territory Electricity Market (NTEM), and the North Western Interconnected System (NWIS)) operate under different rules to the NEM and as such will not be

How to cite this book chapter:
Poulter, H. 2020. Australia: from central electricity to solar/storage systems. In: Burger, C., Froggatt, A., Mitchell, C. and Weinmann, J. (eds.) *Decentralised Energy — a Global Game Changer.* Pp. 25–46. London: Ubiquity Press. DOI: https://doi.org/10.5334/bcf.c.

covered here. The NEM is governed and regulated by a central federal system. This section will give a brief overview of the governance institutions of the NEM and its market operation.

Federal departments

Council of Australian Governments

The Council of Australian Governments (COAG) was initially established in 1991 to drive microeconomic reform in the energy, communications, transport, and water industries (AEMC 2014b), and to replace the Special Premiers Conferences, a similar body who met at frequent intervals to discuss matters of Commonwealth importance. The members are the Prime Minister, State and Territory Premiers, and Chief Ministers and the President of the Australian Local Government Association.

The reform of the electricity industry (the goal of which was to separate policy and regulation from industry; the restructuring of industry; and to introduce competition) resulted in the commencement of the NEM in eastern Australia in 1998 and the Wholesale Electricity Market (WEM) in southern Western Australia in 2006. Since its formation COAG's role has been to promote reforms that are of national significance and/or need the co-operation of all the state and territory governments. These reforms include themes for economic and social participation, national economic competition, sustainable living, health, and reducing the disadvantages of the indigenous people.

Council of Australian Governments Energy Council

The COAG Energy Council was established in December 2013, replacing the Standing Council on Energy and Resources (SCER) and consists of the energy and resource ministers from the Commonwealth, state and territories of Australia, and New Zealand. The council was established in response to a need for energy reforms within both the gas and electricity markets, which included energy efficiency and productivity and energy security, whilst also promoting the competitiveness of Australia's mineral and energy resources, as shown in Figure 5 (COAG Energy Council 2016).

Energy Security Board

The Energy Security Board (ESB) was established in August 2017 following a review of security and reliability of the NEM (Finkel et al. 2017). The ESB consists of the CEO from each of the Australian Energy Market Commission (AEMC), the Australian Energy Regulator (AER), and the Australian Energy Market Operator (AEMO) and an independent Chair and Deputy Chair. The function of the ESB is to provide coordination of the implementation of the reforms put forward by the Finkel Review. The ESB will also 'provide whole of system oversight for energy security and reliability to drive better outcomes for consumers' (COAG Energy Council 2017). The ESB reports directly to the COAG Energy Council.

Australian Energy Market Commission

The AEMC was established in 2005 with the objective to '*promote efficient, reliable and secure energy markets which serve the long-term interests of consumers*' (AMEC 2016). Their function is to review on possible reforms to the current regulatory and market arrangements for both gas and electricity and advise the Energy Council, in conjunction with the AER and the AEMO. This includes managing rule change requests under the National Electricity Rules (NER) and the National Electricity Retail Laws (NERL).

Australian Energy Market Operator

The AEMO is the independent energy markets and power systems operator. It is responsible for the wholesale energy markets (electricity and gas) and management of the NEM. They are a limited company with operating costs recuperated through market fees (AEMO 2016a).

Australian Energy Regulator

The AER regulates the wholesale market of the NEM in compliance with NEM legislation and rules. It works in conjunction with, and is funded by, the Commonwealth with staff, resources and facilities provided from the Australian Competition and Consumer Commission (ACCC). They are the economic regulator of the energy networks and set a maximum price for network charges in collaboration with the network service providers, which is reviewed every five years. They monitor and enforce compliance with the obligations in the Retail Law, Rules and Regulations and provide a price comparison website for retail markets for those states which are compliant under the NERL, which includes Tasmania (TAS), Australian Capital Territory (ACT), South Australia (SA), New South Wales (NSW), and Queensland (QLD) (AER 2009).

Energy Consumers Australia

Energy Consumers Australia (ECA) is a not-for-profit advocacy set up by the COAG Energy Council in 2015 and funded through AEMO to represent the long-term interests of residential and small businesses within the NEM and to provide advice on energy issues to these consumers and represent their views to the Council. They work in conjunction with stakeholders from the energy industry, energy ombudsman, government and market bodies, the research community and media (ECA 2016).

National Electricity Market

The National Electricity Market is the wholesale electricity market covering the eastern coast of Australia. It is governed by the National Electricity Rules (NER) as set out by the regulatory framework of the NEL. It has over 100 participants including generators (the majority taken by coal power plants, transmission network service providers (TNSPs), distribution network service providers (DNSPs), and market customers).

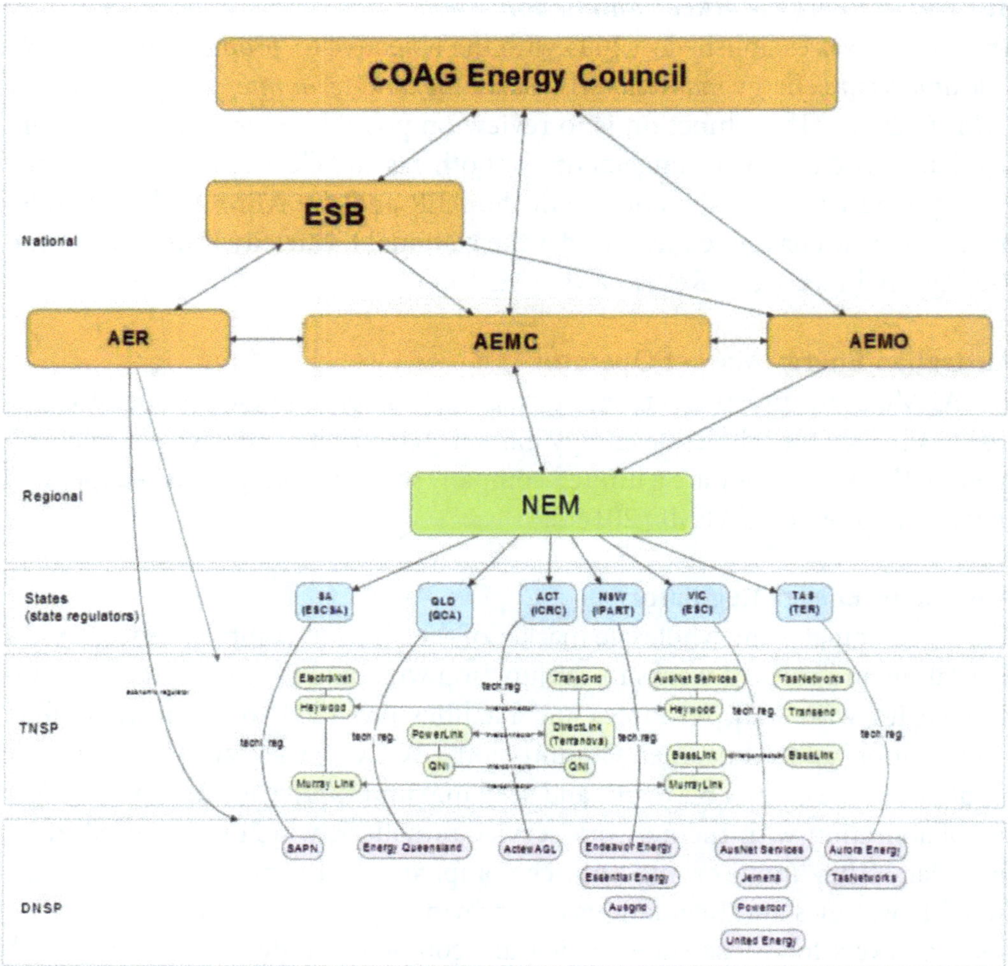

Figure 5: Electricity network governance structure of Eastern Australia.
Source: Author based in information sourced in this chapter.

The NEM is a transmission and distribution grid covering 5 interconnected state-based networks (South Australia (SA), Tasmania (TA), Victoria (VIC), New South Wales (NSW) (including the Australian Capital Territory (ACT)), and Queensland (QLD) covering a distance of approximately 5000 km and is operated by AEMO. It is an energy-only gross pool market with 5 minute spot prices averaged for half-hourly periods in each of the six states. Bids are taken for each of the 6 trading intervals and the averaged spot price is received by all generators dispatched in the trading interval. A recent rule change, which will take effect in 2021, is to reduce the settlement period from 30 minutes to five minutes. This change is to provide a better signal for investment into the market for new technologies and business models, such as batteries and demand response. It has been suggested that it will also reduce the 'gaming' of the half-hourly settlement by gas peaking generators and should therefore reduce prices (McConnel & Sandiford 2016; Parkinson 2016). Risks in price volatility are reduced using hedges, options, and futures contracts (AEMO 2016c).

2.2.3 *Australia's renewable energy resources*

Australia's renewable energy resources comprise of solar, wind, hydro, ocean, and bioenergy. Australia has the highest solar irradiation of any continent and average wind speeds in the southern part of the continent in excess of 6 m/s (the minimum average wind speed used in assessing wind farm potential). Due to the presence in the south of a westerly wind known as the 'roaring forties', wind farms in South Australia have reported capacity factors of, on average, 33 per cent. The excellent solar resource is able to produce capacity factors of around 30 per cent in summer and 15 per cent to 18 per cent in winter (AEMO 2016c). The high potential of these resources has meant that *new* renewable installations (Tasmania installed its first hydropower station a century ago) have been predominantly wind and solar.

By the end of June 2018, installed capacity, including operating and under-construction facilities of large-scale wind and solar had reached 7565 MW with a further 774 MW of projects signing power purchase agreements (Clean Energy Regulator 2018). Small-scale PV (systems under 100 kW) split into two categories currently have:

- commercial systems (10–100 kW) of 1063 MW installed capacity
- residential systems (under 10 kW) of 5556.8 MW of installed capacity (APVI 2018a).[4]

In 2017–18 Australia had 44.8 GW of electricity generation capacity (AER 2019). Small-scale distributed solar now makes up 14.8 per cent of installed capacity.

2.2.4 *Distributed Renewable Generation in Australia*

The quality of the solar resource, falling prices for PV and storage, and high electricity prices have made self-generation an excellent proposal for residential and business consumers in Australia. Since 2008, the NEM states have offered their own version of feed-in-tariffs, and this, in conjunction with the Small-scale Renewable Energy Scheme, has led to the current size of the domestic market. This section will look at this in more detail.

Solar schemes in the NEM

In 2008 as part of their own renewable energy targets, and to encourage early adopters at a household and commercial level, each of the Australian states introduced their own incentives for small-scale solar (<100 kW). These state designed schemes were a Solar Bonus Scheme (SBS) in New South Wales

[4] Animation for PV installations from 2007–2018 can be found at http://pv-map.apvi.org.au/animation.

Table 2: FiT rates for NEM states (2017).

STATE	Scheme	Rate c/kWh	Max size
Queensland	South-eastern QLD: no minimum Rural QLD: mandatory minimum	Based on retailer competition 6–8	5 kW
New South Wales	Recommended benchmark range for retailers	11.6–14.6	Depends on retailer
Australian Capital Territory	No minimum	Depends on retailer: currently 6-8	n/a
Victoria	Mandatory minimum	11.3	<100kW
South Australia	No minimum	Depends on retailer: currently 6–12	First 45 kWh per day
Tasmania	Set rate	7	10 kW single phase 30 kW three phase

Source: Energy Matters (2017).

(NSW) and a Premium feed-in-tariff (FiT) in Tasmania (TAS) and the Australian Capital Territory (ACT). The SBS and Premium FiT paid between 44c per kWh and 60c per kWh dependent on state regulated tariffs for *all* PV generated electricity. South Australia (SA) and Queensland (QLD) had a net FiT which paid 44–60 per kWh for electricity that was fed back into the grid. All state schemes were initially open to unlimited size of domestic system but then curtailed to 5 kW. In 2012, the schemes closed to new entrants and the tariff reduced to around 8c per kWh. Those whose systems were registered and installed before the 2012 deadline would receive the initial Premium or Net FiT or SBS until 2028. Future FiTs will be set and paid by the retailer with no minimum requirement but with the states able to set minimums or provide benchmark ranges as a guide for retailers if they wish (Table 2).

The small-scale renewable energy scheme

In 2011 the Small-scale Renewable Energy Scheme (SRES) was introduced as a measure to encourage individuals and small businesses to invest in eligible renewable energy (RE) systems (CER 2018). The scheme will run until 2030. SRES certificates are produced for new renewable generation by solar PV, wind, or hydro and for the energy displaced by a solar water heater or heat pump over the course of a designated period. The certificates can be generated by the owner of the eligible technology or by an installer who has been assigned the right to generate the certificate by the owner. As one certificate is equal to 1 MWh, small-scale renewable energy (RE) installers can aggregate smaller RE systems to reach the 1 MWh target and pass on savings to the customer. The certificates are sold to energy retailers, who are required to surrender an amount of certificates each year, through a spot market or through a Government Clearing House (RET 2018). The value of the certificates is dependent on the spot market price, currently Aus$35.50/MWh. Alternatively, certificates can be traded through the government Clearing House at Aus$40/MWh. Certificates will only be traded through the Clearing House if there are no certificates available through the spot market. The added benefit of the scheme is that it gives the AEMO visibility of the location and density of behind-the-meter RE which allows for better forecasting for grid demand.

The rise of DER in Australia

During the years between 2010 and 2012, Australia saw a huge uptake of distributed solar PV with 900,000 of the current 1.7 million systems being installed in those three years (Figure 6).

This led to the highest percentage penetration rates of domestic PV systems worldwide, double that of Belgium and three times that of Germany and the United Kingdom (AEC 2016), as shown in Figure 7.

The result of this uptake has meant that in SA there has been a shift in the time of operational peak demand as the capacity of residential PV systems

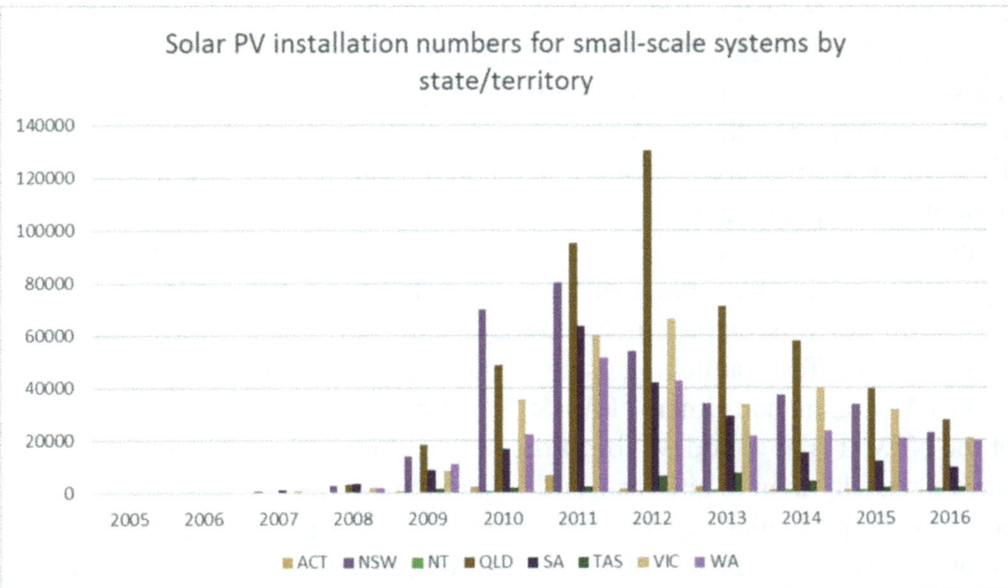

Figure 6: Number of domestic solar PV systems installed by year in the NEM states.

Source: Clean Energy Regulator (2016).

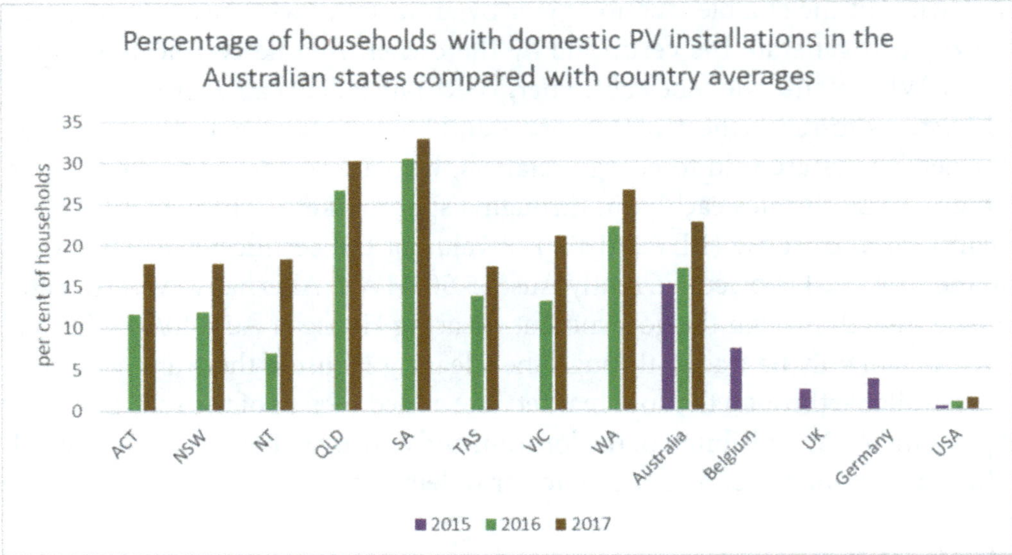

Figure 7: Comparison of state/territories of Australia at year end 2016 (Clean Energy Regulator 2016), end of May 2017 (Roy Morgan Research 2017) and world averages in 2015.

Source: AEC (2016).

(679 MW) reduced the daytime peak demand profile by almost 10 per cent in 2015–16 (AEMO 2016c), producing what is now being called a 'duck-curve'. In its *National Electricity Forecasting Report* (AEMO 2016a), AEMO (the market operator) predicts a reduction and flattening of consumption for grid supplied electricity due to further uptake of PV from both business and residential customers, and an increase in the energy efficiency of households and appliances. It also recognises that maximum demand for generated power will continue to shift to both later in the day and year, towards the Australian winter evenings. This shift has led to a reduction for demand in baseload power but an increase in demand for peaking plants.

Australia has been the obvious choice for the introduction of residential storage due to its dominance of the rooftop PV market. In 2016, companies competing for this market included Tesla, Enphase, GCL Poly, Sonnen, and RedFlow, with Tesla launching its new Powerwall 2.0 energy storage device at a cost to Australian consumers of Aus\$0.23 kWh^{-1} (Peacock 2016). In some states of the NEM it is now, and will be in the future, cheaper to install, generate, and store electricity than to buy retail and, dependent on pricing structures, be grid connected e.g. retailers are now charging a standing rate for meters which is between 80 and 150 c per day or between Aus\$292 and Aus\$547.50 per year – with the average price for installing a 5 kW PV array plus storage cAus\$16,000 (Mountain 2016).

In Figure 8, the cost of the PV/storage system in the NEM states has been calculated for a 10 year lifetime (the average lifespan of a battery system), however,

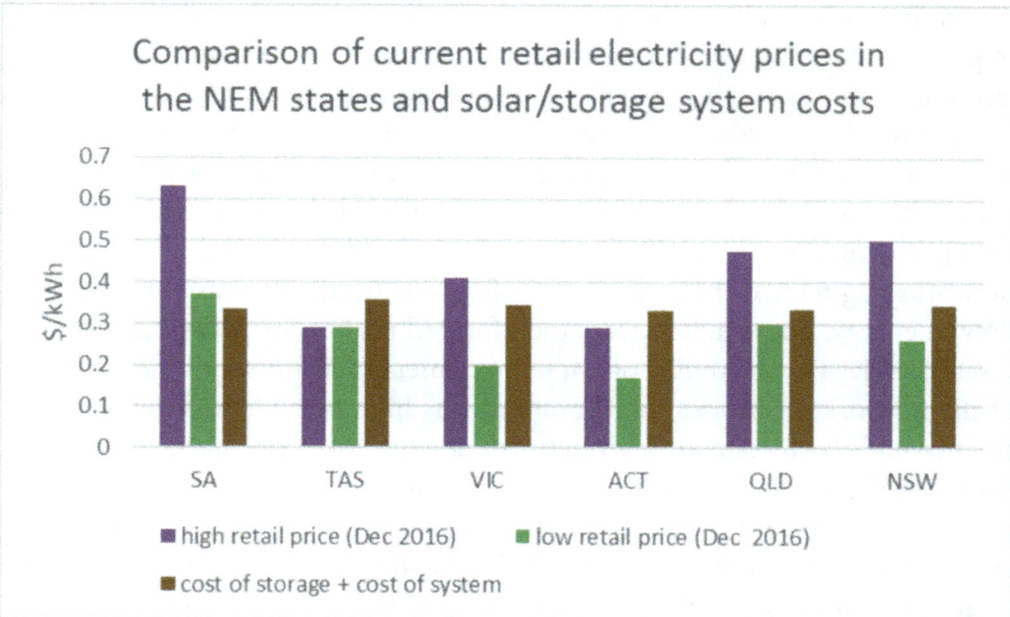

Assuming lifespan of 10 years for both (after 10 years costs drop to Aus$0), costs for PV taken from solar choice.net.au and solar resource and household electricity use for each of the capital cities calculated for each of the NEM states. As it is assumed that these households would not use the grid, no allowance has been made for FiT payments

Figure 8: Cost of a solar/storage system.

Source: Australian Government (2017), Peacock (2016), Solar Choice (2017).

if costs were to be spread over a 20 year lifetime, QLD and TAS would also see DER costs cheaper than retail prices. These figures do not include government FiT schemes in order to indicate the cost of the system assuming the customer wanted to disconnect from the grid. In AEMO's *South Australian Electricity Report* (AEMO 2016b), with figures based on information available until July 2016, expected uptake of combined PV and storage systems was predicted to begin slowly and not see growth until after 2020. It also predicted that retrofit of storage systems would be uneconomical. However, nationally in 2016, 6750 storage units were installed, up from the previous year's figure of 500, and this with no government or policy support. It was estimated that in 2017, due to the late arrival of the Powerwall 2.0 and new technology that would ease the retrofitting and installation of storage units, that Australia would install as many as 20,000 domestic scale units (Morris 2017). A report from the Clean Energy Council (Clean Energy Council 2018) using figures from SunWiz has now stated that the current number of installations (as at the end of 2017) stands at 28,000 with 12 per cent of installations in 2017 also incorporating storage, up from 5 per cent the previous year.

After the commencement of the NEM the real term cost of retail electricity fell until 2007. In 2010–12 there was steep rise with various factors combining causing this rise i) a growth in peak demand requiring new infrastructure and generating capacity, ii) retirement of old coal thermal plant, and iii) switching to gas generation for peaking plants and baseload power at a time when global gas prices were increasing (Simshauser & Laochumnanvanit 2011).

Forecasts had predicted an increase in electricity demand which would require new infrastructure and generating capacity. In fact demand fell which meant that electricity companies invested in unneeded infrastructure (Saddler 2017). The incorrectly forecast increase in demand meant that the electricity networks augmented their power lines for an expected increase in a one-way flow of energy. These investments then formed part of the networks regulated asset base (RAB). The return on these investments accounts for more than half of the network businesses total revenue. This high revenue and hence a continuing high RAB leads to high network charges for consumers. In Australia the network charges currently make up around 40 per cent of customers' bills.

Old coal thermal power plants were retired and replaced with combined cycle gas thermal power plants (CCGT). At this time there was an increase in global gas prices which led to the development of 3 LNG export projects in Queensland. These developments limited the availability of local gas for the domestic market and caused wholesale gas prices to rise, affecting the cost of electricity supplied by gas generation in the NEM (Oakley Greenwood 2016) and therefore increasing prices for consumers. Forecasts predict more increases for residential electricity prices in 2020, between 15 and 20 per cent higher than 2016 levels.

Future price rises have been attributed to a reliance on gas for future electricity supply and an unexpected reduction in electricity demand leading to over-investment in generation capacity (Reed 2016). In particular South Australia (SA) and Queensland (QLD) are more reliant on gas generation for baseload power. It is forecasted that commercial and industrial users could see increases of 20–40 per cent compared to 2016 levels by 2037 (Parisot & Nidras 2016).

For domestic customers the retail cost of electricity is made up of two separate charges, a daily supply charge, between 80 and 150 c per day dependent on state and supplier, and the volumetric charge, which includes costs for distribution, transmission, retail, and wholesale costs, and environmental policies. On 1st July 2016 the network prices in the ACT and NSW increased, but decreased in QLD, SA, and TAS (Duffy & Johnston 2016) therefore the rise in retail prices seen in the latter states could be attributed to an increase in generation costs and/or environmental policies, or due to the reduction in volume of sales from customers.

Following the continuing price rises the Australian Competition and Consumer Commission (ACCC) released a report on retail pricing in the NEM (ACCC 2018). The report recognises that competition in the NEM is not working and that the current approach to policy, regulatory design, and promotion of competition is not working in the best interests for consumers. The report has 56 recommendations around four broad areas:

1. Boosting competition in generation and retail.
2. Lowering costs in networks, environmental schemes and retail.
3. Enhancing consumer experiences and outcomes.
4. Improving business outcomes.

2.2.5 Disruption within the electricity industry

There has been both climatic and technological disruptions within the electricity industry in Australia. Climatic disruption resulting in blackouts (as a result of extreme heatwaves), and technological disruption – the unprecedented penetration of DER, causing some commentators to suggest that Australia may be soon in the grips of a utility 'death spiral'.

Box 1: Climatic events

On the 28th September 2016 storms in northern South Australia, including tornados with wind speeds in the range of 190–260 km/h caused a state-wide blackout. The update report (AEMO 2017b) from AEMO and also a recent review by the Chief Scientist, Alan Finkel (Finkel et al. 2017) have recommended that value services for Frequency Controlled Ancillary Service (FCAS) and System Restart Ancillary Services (SRAS) from distributed energy and other storage technologies should be investigated.

In addition, on February 8th 2017, a heatwave with temperatures of 41.6°C left 90,000 homes (AEMO 2017c) in South Australia without power as AEMO committed to load shedding due to failures of its fossil fuel generators. The demand forecasts they had received had not anticipated the cooling demand for the period and the reduced wind power due to a drop in wind speed. A 165 MW gas plant was 'unavailable' and unable to start-up in the time requested, needing 4 hours notice, leaving homes without power in the evening on one of the hottest days of the summer.

The 28th September 2016 events in detail:

- The northern South Australia Storms, including tornados with wind speeds in the range of 190–260 km/h, caused multiple network faults and downed 3 major transmission lines;
- the resulting voltage disturbances caused settings on the wind generators to reduce their output;

- a necessary increase in imports through the Heywood interconnector from Victoria caused a loss of synchronism and led to one of the interconnectors tripping;
- the FCAS capability of the gas generators – which should have stopped a collapse of system frequency was unable to respond at the required speed;
- the SRAS – which also should have stopped a collapse of system frequency – were unable to start due to unexplained failures;
- this led to a collapse of electricity system frequency.

These two events happened in an Australian state which has over 30 per cent of households with rooftop PV and in which installing combined domestic scale solar and storage is cheaper than buying grid electricity. As can be seen from Figure 12 there has been a 5.5 per cent rise in the number of domestic PV installations in the first half of 2017 which has been attributed in some way to these events.

The death spiral effect

A death spiral is a positive feed-back loop. This occurs when, due to rising network charges within electricity prices, consumers switch to their own on-site generation and/or leave suppliers. This then leaves the initial network company supplier with either less customers or less kWhs from which to recoup costs thus having to raise prices, and so on, until demand eventually collapses. Within energy this phenomenon has been recognised as a utility death spiral.

The term 'utility death spiral' saw a re-emergence recently in Australia. The term was originally coined with the new era of competition as energy markets opened up and the rising costs of some utilities meant that consumers could switch to cheaper alternatives/sources (Costello & Hemphill 1990). Then, it was much more a conceptual argument – a threat – to ensure policies which did not lead to a death spiral. A perfect storm of requirements would be needed to induce such a spiral – inflexible pricing structures, large defections and the utilities unable to change their behaviour (Graffy & Kihm 2014). Increasingly, the possibility of consumers generating and storing their own power and adding in a new type of disruptive competition has added to the potential drivers for a death spiral.

As can be seen from the current situation in Australia, the combination of a high percentage of householders already with PV installed, access to cheap storage, and high priced retail electricity means that the Australian electricity markets could be experiencing the beginnings of a death spiral which could threaten the existing incumbents.

Utility costs – those for network charges currently make up 40–50 per cent of the Australian retail price (AEMC 2016), in comparison to the United Kingdom, where network costs are approximately 27 per cent (Ofgem 2017). There are high retail costs currently and further rises expected in the future, which will invariably mean that the rate of return for the customers installing DER systems will be favourable. In response to this, and the need to reduce the percentage of the retail charge for network costs, retail companies include a standing charge (average Aus$400 per year) for domestic customers as well as the volumetric charge. It is now affordable for domestic customers to generate and store the majority of their electricity needs (Mountain 2016). If a larger and larger percentage of people install DER, so a smaller proportion of grid-generated electricity is needed. Should storage costs continue to fall, and domestic generators not given value for their contribution to decarbonising the energy system, disconnecting from the grid may become a viable choice for Australian householders. This suggests that, if Australia wishes to limit the possibility of a death spiral for its utilities, it will need to create value for DER for both the utilities and the consumer. Certainly, continuing in its current position is widely recognised as not an option.

2.2.6 The future of the Australian electricity system

Australia is in a unique position as high electricity prices, falling costs for DER, and issues concerning the reliability of electricity supply has caused, as of 2017, just over a quarter of households to install solar PV. There has also been significant rises in larger (50–100 kW) installations for business and small commercial properties (APVI 2018b). The threat of a possible death spiral for utilities and the increase in frequency and ferocity of climatic events, as predicted for anthropogenic climate change, have highlighted the need for a more reliable and flexible grid; and a market which rewards DER – something that Australia does not have currently. AEMO (the market operator) has recently appointed Audrey Zibelman, ex Commissioner of the New York (NY) Public Service Commission in New York state (NYS), United States, as the new Head of AEMO. Ms Zibelman was the driver of the NYS's progressive market and network regulation known as the NY Reforming the Energy Vision (REV) (Mitchell 2016), and one can expect therefore similar ideas as those set out in the NY REV to occur in Australia. In a recent interview (ABC Radio 2017) Audrey Zibelman explained how Hurricane Sandy caused New York to rethink its energy strategy. She also talked of the implementation of a solar-based micro-grid in Brooklyn and said she hopes to see such schemes, of which there are a few being trialled within the Australian electricity markets, soon being an integral part of the new energy system in Australia.

In December 2017, following AEMO's (and particularly Ms Zibelman's) support of trialling demand response to alleviate extreme demand peaks during high summer temperatures (AEMO 2017a), the AER commenced a Demand Reduction Incentive Scheme and Innovation Allowance Mechanism (AER 2017). The Incentive is targeted at the distribution companies to use non-network solutions to provide reduction in demand and the Mechanism provides a small amount of funding for R&D in demand management projects. There is also an allowance (Aus$200 K) to help with the roll out of any innovation projects. One trial has been for a community-based pilot project, Power Changers (Jemena 2018). The trial is being led by Jemena, one of the Victorian distribution companies. The project uses an app to encourage users to complete challenges to reduce their demand. If the challenges are completed, they receive reward points. These reward points are then collected and converted into a monetary reward given to an organisation chosen by the community, such as a school.

The NEM also has the problem of many rural customers at the end of long, thin distribution wires. In some cases these wires can extend, over terrain with a high bush fire risk, for 100 km to serve just a handful of customers. The costs of maintaining this wire are exorbitant and, as all network costs are spread over all consumers; microgrids would enable a more reliable service for rural customers and an overall reduction in customer bills. Although an excellent solution, there are problems arising from this for a privately owned, competitive market such as the NEM. The question currently being examined is how to introduce competition for retail, generation, and storage in micro-grids whilst ensuring consumer protections

In the NEM the big 3 'gentailers' – Origin Energy, AGL, and EnergyAustralia – have been encouraging the use of battery systems by offering PV and battery packages through their retail services. Although this seems contrary to what would be expected from fossil fuel generating companies, it shows that some of the Australian generators are accepting that the future energy system will need to include a large proportion of DER and are therefore already deciding to enter the field.

As well as offering a residential and commercial storage package, AGL are trialling a Virtual Power Plant (VPP) in Adelaide, SA (AGL 2017). The VPP is made up of 1000 domestic solar PV and battery systems which are aggregated and so can be seen on the network. These can then be used for network services such as peak demand reduction and frequency control. This will help to create resilience for the network and reliability for the consumer. A similar scheme is also underway in Bruny Island, Tasmania. Reposit Power in collaboration with the Australian Renewable Energy Association (ARENA), TasNetworks, and academia in Melbourne and Tasmanian Universities, have set up a VPP to reduce the islands reliance on back-up diesel generation (Reposit Power 2017).

In 2017 in SA the previous Labor government and Tesla began a trial of a VPP of 50,000 homes. The initial trial installed 5 kW of rooftop solar and a 15.3 kWh Tesla Powerwall battery on 100 housing association homes. This trial is the first stage which will see another 1000 homes with solar and batteries installed in the second stage. The original trial was then to extend this to a total of 50,000 homes in a third stage. However, after a recent change of government it is expected that this will change to the election promise of Aus$100 m of grants towards domestic battery storage, an estimated 40,000 homes. The new government has still shown an interest in the Tesla proposal and is also interested in other battery companies such as Sonnen and is promising that SA will still lead the way in renewable energy transformation (Parkinson 2018).

Due to the rise in DER there has also been the question of tariff reform to reduce the 'death spiral' effect and increase visibility of DER. One idea being spoken about is to change from a kWh charge to a kWp (kilowatt peak) charge. This tariff would be similar in effect to a mobile phone data plan. The customer would choose a plan based on their peak kilowatt usage e.g. not going above 3 kW of demand at any one time. If they go over this limit they would have to pay a fee. They would then be given the option to increase their plan if they thought that this would be something that would happen often e.g. they install an air conditioner which uses energy when their PV system is not generating or charge an electric vehicle at peak times. This would then give the customer the option of (1) changing their consumption patterns, (2) changing to a new plan, or (3) buying energy storage dependent on the economics and/or customer preferences. The idea behind this is that high costs for the consumer come from the need for peaking generators and distribution capacity to cover these peaks. By customers committing to a peak level this will give the distribution companies and the market operator visibility of where, and how much, generation and network capacity is needed.

The Finkel Review

Following the events of September 2016 the federal government undertook a review of the governance of the NEM. The review was undertaken by the Office of Australia's Chief Scientist – Dr Alan Finkel. The review 'An Independent Review into the Future Security of the National Electricity Market: Blueprint for the Future' (Finkel et al. 2017) made recommendations on the themes of:

- **increased energy security within the NEM** for increased penetration of variable renewable resources, including valuing frequency response, synthetic inertia, demand response, and voltage control and also cyber security due to the increase of IT services within the system;
- **policy stability** with recommendations for a long ranging Clean Energy Target (CET) which would see certificates issued for all types of generation with more certificates issued for the least polluting technologies;
- **efficiency within the gas markets** to ensure that electricity generators are able to maintain reliability of supply;
- **improved system planning** to include a transmission and distribution plan to recognise areas of future economically viable VRE penetration, also a review of regulation to remove the incentives for networks to prioritise over non-network solutions;

- **rewarding consumers** including the facilitation of a DER market and a change in role for the distribution networks to provide a platform for new technologies;
- **stronger governance** to include the establishment of an Energy Security Board (ESB) to oversee the implementation of the plan and to be a single point of responsibility and accountability between market institutions and the Energy Council. This area will also review the rule-change process to accommodate the rapidly changing energy market.

In total there were 50 recommendations contained within the main points above. In July 2017 the review was presented to the COAG Energy Council for approval and 49 out of the 50 recommendations were approved. The council felt that the Clean Energy Target would need further consideration.

The Energy Security Board was appointed in September 2017 and announced plans at the end of 2017 for a National Energy Guarantee (NEG). The NEG will combine a Reliability Guarantee and an Emissions Guarantee. In order for the policy to take effect, it will need approval from all members of the COAG Energy Council. Following consultation in April 2018 the final version of the NEG will be published in August 2018 (note that, following a review by the Energy Council, the NEG has not been approved and Australia currently has no energy policy post 2020).

Following the Finkel Review recommendations AEMO have released their Integrated System Plan (ISP) (AEMO 2018). The plan recognises that there are fundamental changes happening within the Australian energy system:

- flattening of grid demand due to the rise in behind the meter DER and energy efficiency (even assuming a rise in EV ownership);
- the coal power stations that currently provide the majority of power are due to be retired in the next 20 years;
- the costs and capabilities of new supply resources have changed significantly and are expected to do so in the future;
- renewables, storage technologies, and flexible gas-powered generation are expected to be the core components to a low cost and reliable energy future.

In order to incorporate renewables, both at grid scale and at domestic scale, the Integrated System Plan makes recommendations for increased interconnection between the states, investment into Renewable Energy Zones (REZ), and coordination of behind the meter DER. AEMO and Energy Networks Australia (ENA) have also released a consultation paper on 'how best to transition to a two-way grid that allows better integration of Distributed Energy Resources for the benefit of all customers' (AEMO & ENA 2018). AEMO and the ENA have given possible frameworks for integrating DER:

a. a single integrated platform where the market operator provides a central platform in which all distribution level actors are able to participate;

b. a two-step tiered platform in which the DNOs are responsible for optimisation and DER dispatch within their own network areas;
c. an independent distribution service operator (DSO) or AEMO optimising distribution level dispatch which would involve either an overarching iDSO or multiple iDSOs at the individual network level.

Within the report the agencies recognise the value that DER could bring if given an effective framework in which to operate. This is in comparison to a lack of coordination which would increase costs for everyone.

2.2.7 Conclusion

2018 saw the beginning of many changes to the Australian electricity system with the COAG Energy Council agreeing to implement 49 out of the 50 recommendations made by the Finkel review, in particular the replacement of the current Renewable Energy Target with the controversial National Energy Guarantee. For DER the Finkel review includes giving value to customers for demand management, changing the role of the distribution networks to facilitate a DER market and a review of network regulation. Should the NEM governance institutions follow the advice given within the review, then this will help to avoid new network costs and promote alternative non-network solutions which will enable *all* customers to benefit from a flexible, low-emission electricity system.

Australia is experiencing particular governance issues due to the disruption caused by the rapid increase in DER and the effect of this on a huge interconnected transmission grid such as the NEM. A new system operation which offers a cost-effective solution to both these challenges can be provided by DER, and we can expect – given the Finkel Review, the acceptance of DER by some of the utility companies, and Audrey Zibelman's new appointment – that this is where the momentum is within Australian system operation.

Author's note

A small amount of the above chapter was previously published by the author on the IGov New Thinking for Energy blog, found at http://projects.exeter.ac.uk/igov/new-thinking-tales-of-the-unexpected/

2.2.8 References

ABC Radio. (2017), *Audrey Zibelman interview.* Retrieved July 18, 2017, from http://mpegmedia.abc.net.au/rn/podcast/2017/04/sea_20170401_0745.mp3

ACCC. (2018), *Restoring electricity affordability and Australia's competitive advantage: retail electricity pricing inquiry-final report, Canberra.* Retreived

July 23, 2018, from https://www.accc.gov.au/system/files/Retail Electricity Pricing Inquiry—Final Report June 2018_Exec summary.pdf

AEC. (2016), *Renewable energy in Australia – how do we really compare?, Australian Energy Council*. Retreived February 10, 2019, from https://www.energycouncil.com.au/, accessed February 10, 2019

AEMC. (2014b), *National electricity market: a study in successful microeconomic reform*. Retrived from http://www.aemo.com.au/About-the-Industry/Energy-Markets/National-Electricity-Market.

AEMC. (2016), *2016 residential electricity price trends, Sydney*. Retrieved August 31, 2017, from http://www.aemc.gov.au/getattachment/be91ba47-45df-48ee-9dde-e67d68d2e4d4/2016-Electricity-Price-Trends-Report.aspx

AEMO. (2016a), *National electricity forecasting report, Australian Energy Market Operator*. Retrieved from http://www.aemo.com.au

AEMO. (2016b), *South Australian electricity report, Australian Energy Market Operator*. Retreived from http://www.aemo.com.au

AEMO. (2016c), *South Australian renewable energy report, Australian Energy Market Operator*. Retrieved from http://www.aemo.com.au

AEMO. (2017a), *ARENA and AEMO join forces to pilot demand response to manage extreme peaks this summer*. Retrieved March 13, 2018, from https://www.aemo.com.au/Media-Centre/ARENA-and-AEMO-join-forces-to-pilot-demand-response-to-manage-extreme-peaks-this-summer

AEMO. (2017b), *Black system South Australia 28 September 2016, Melbourne*. Retreived July 18, 2017, from https://www.aemo.com.au/-/media/Files/Electricity/NEM/Market_Notices_and_Events/Power_System_Incident_Reports/2017/Integrated-Final-Report-SA-Black-System-28-September-2016.pdf

AEMO. (2017c), *System event report South Australia, 8 February 2017, Melbourne*. Retrieved July 18,2017, from https://www.aemo.com.au/-/media/Files/Electricity/NEM/Market_Notices_and_Events/Power_System_Incident_Reports/2017/System-Event-Report-South-Australia-8-February-2017.pdf

AEMO. (2018), *Integrated system plan for the national electricity market, Melbourne*. Retreived July 23, 2018, from http://www.aemo.com.au/-/media/Files/Electricity/NEM/Planning_and_Forecasting/ISP/2018/Integrated-System-Plan-2018_final.pdf

AEMO & ENA. (2018), *Open energy networks: consultation on how best to transition to a two-way grid that allows better integration of distributed energy resources for the benefit of all consumers*. Retrieved July 23, 2018, from https://www.aemo.com.au/-/media/Files/Electricity/NEM/DER/2018/OEN-Final.pdf

AER. (2009), appendix: energy market reform.

AER. (2017), *Demand management incentive scheme, Melbourne*. Retreived Januart 17, 2018, from https://www.aer.gov.au/system/files/AER – Demand management incentive scheme – 14 December 2017.pdf

AER. (2019), *Generation capacity and peak demand|Australian Energy Regulator.* Retrieved from https://www.aer.gov.au/wholesale-markets/wholesale-statistics/generation-capacity-and-peak-demand

AGL. (2017), *Virtual power plant – South Australia.* Retreived August 5, 2017, from https://aglsolar.com.au/power-in-numbers/

APVI. (2018a), *Mapping Australian photovoltaic installations, Australian Photovoltaic Institute.* Retreived July 23, 2018, from http://pv-map.apvi.org.au/historical#4/-26.67/134.12

APVI. (2018b), *PV postcode data, Australian Photovoltaic Institute.* Retreived March 13, 2018, from http://pv-map.apvi.org.au/postcode

Australian Government. (2017), *Find an energy plan, Energy made easy.* Retreived March 13, 2018, from https://www.energymadeeasy.gov.au/

Clean Energy Council. (2018), *Clean energy Australia, Melbourne.* Retreived from https://www.cleanenergycouncil.org.au/resources/resources-hub/clean-energy-australia-report

Clean Energy Regulator. (2016), *Postcode data for small-scale installations.* Retreived January 10, 2017, from http://www.cleanenergyregulator.gov.au/RET/Forms-and-resources/Postcode-data-for-small-scale-installations#Smallscale-installations-by-installation-year

Clean Energy Regulator. (2018), *Large-scale renewable energy target market data.* Retreived July 23, 2018, from http://www.cleanenergyregulator.gov.au/RET/About-the-Renewable-Energy-Target/Large-scale-Renewable-Energy-Target-market-data

COAG. (2016), About COAG | Council of Australian Governments (COAG) Retrieved November 21, 2016 from http://www.coag.gov.au/about_coag

COAG Energy Council. (2017), *Energy security board.* Retreived March 13, 2018, from http://www.coagenergycouncil.gov.au/energy-security-board

Costello, K.W., Hemphill, R.C. (1990), Competitive Pricing In The Electric Industry*. Resour. Energy 12, 49–63.

Dufty, G. & Johnston, M.M. (2016), *The NEM – A hazy retail maze, Melbourne: Alviss Consulting.* Retreived February 10, 2019, from https://alvissconsulting.com/the-national-energy-market-a-hazy-retail-maze/

ECA. (2016), Mission and vision. Retreived September 9, 2019, from https://energyconsumersaustralia.com.au/about-us/mission-and-vision

Energy Matters. (2017), *Australian solar feed-in tariff information – Energy Matters.* Retreived August 2, 2017, from http://www.energymatters.com.au/rebates-incentives/feedintariff/#northern-territory

Felder, F.A. & Athawale, R. (2014), The life and death of the utility death spiral. *Electricity Journal, 27*(6), 9–16.

Finkel, A., Moses, K., Munro, C., Effeney, T. & O'Kane, M. (2017), *Independent review into the future security of the National Electricity Market – blueprint for the future. Canberra.* Retreived August 5, 2017, from see http://www.environment.gov.au/system/files/resources/1d6b0464-6162-4223-ac08-3395a6b1c7fa/files/electricity-market-review-final-report.pdf

Graffy, E., Kihm, S. 2014, Does disruptive competition mean a death spiral for electric utilities? 35, 1–31.

Horizon Power. (2017a), *Install solar|Buyback eligibility|Horizon Power*. Retreived July 20, 2017, fromhttps://horizonpower.com.au/solar/eligibility-to-install-and-buyback-schemes/#for-home

Horizon Power. (2017b), *Onslow power project. Onslow Power Project*. Retreived August 5, 2017, from https://horizonpower.com.au/our-community/projects/onslow-power-project/

Horizon Power. (2017c), *Stand-alone power systems project*. Retrieved August 5, 2017, from https://horizonpower.com.au/our-community/projects/stand-alone-power-systems-project/

Jacana Energy. (2017), *Solar buyback rates*. Retreived August 2, 2017, from https://jacanaenergy.com.au/photovoltaic_pv_solar_systems/solar_buy back_rates

Jemena. (2018), *Power changers pilot*. Retreived March 13, 2018, from http://jemena.com.au/home-and-business/electricity/my-electricity-supply/power-changers-pilot

Laws, N.D., Epps, B. P., Peterson, S.O., Laser, M. S. & Wanjiru, G K. (2016), *On the utility death spiral and the impact of utility rate structures on the adoption of residential solar photovoltaics and energy storage. Applied Energy*. Retreived from http://linkinghub.elsevier.com/retrieve/pii/S0306261916315732

McConnell, D. & Sandiford, P.M. (2016), Winds of change: an analysis of recent changes in the South Australian electricity market. Melbourne Energy Institute

Mitchell, C. (2016), *US regulatory reform: NY utility transformation. IGov: new thinking for energy*. Retreived July 18, 2017, from http://projects.exeter.ac.uk/igov/us-regulatory-reform-ny-utility-transformation/

Morris, N. (2017), *Battery storage_ Is Australia on track to be the world's biggest market? One step off the grid*. Retreived February 8, 2017, from https://onestepoffthegrid.com.au/battery-storage-australia-track-world-biggest-market/?utm_source=RE+Daily+Newsletter&utm_campaign=afd71aa60f-EMAIL_CAMPAIGN_2017_02_08&utm_medium=email&utm_term=0_46a1943223-afd71aa60f-40390089

Mountain, B. (2016), *Tesla's price shock: solar + battery as cheap as grid power. RenewEconomy*. Retreived September 25, 2017, from http://reneweconomy.com.au/teslas-price-shock-solar-battery-as-cheap-as-grid-power-22265/

Murgia, M. & Thomas, N. (2017, March 12), DeepMind and National Grid in AI talks to balance energy supply, *Financial Times*.

Oakley Greenwood. (2016), Gas Price Trends Review.

Ofgem. (2017), *Infographic: Bills, prices and profits*. Retreived August 31, 2017, from https://www.ofgem.gov.uk/publications-and-updates/infographic-bills-prices-and-profits

Parisot, L. & Nidras, P. (2016), Retail electricity price history and projections – Public. Melbourne: Jacobs Australia Pty

Parkinson, G. (2016), *Change market rules, and battery storage will easily beat gas. RenewEconomy.* Retrieved March 13, 2018, from http://reneweconomy.com.au/change-market-rules-and-battery-storage-will-easily-beat-gas-12917/

Parkinson, G. (2017a), *Network limits on solar, storage could accelerate "death spiral." Reneweconomy.* Retrieved April 6, 2017, from http://reneweconomy.com.au/network-limits-on-solar-storage-could-accelerate-death-spiral-65424/

Parkinson, G. (2017b), *States threaten to go it alone on clean energy as Coalition loses plot. RenewEconomy.* Retrieved July 13, 2017, from http://reneweconomy.com.au/states-threaten-to-go-it-alone-on-clean-energy-as-coalition-loses-plot-70162/?utm_source=RE+Daily+Newsletter&utm_campaign=d52c96b021-EMAIL_CAMPAIGN_2017_07_12&utm_medium=email&utm_term=0_46a1943223-d52c96b021-40390089

Parkinson, G. (2018), *SA Liberals vow to continue energy transition, go big in batteries: RenewEconomy.* Retrieved July 23, 2018, from https://reneweconomy.com.au/sa-liberals-vow-to-continue-energy-transition-go-big-in-batteries-93367/

Peacock, F. (2016), *Graph of the day – Tesla Powerwall 2 way ahead of competition on price: Renew Economy.* Retrieved January 10, 2017, from http://reneweconomy.com.au/graph-of-the-day-tesla-powerwall-2-way-ahead-of-competition-on-price-62928/

Reed, T. (2016), Energy prices Part 2: Why is this happening? – Ai Group Blog . Retrieved January 31, 2017 from http://blog.aigroup.com.au/energy-prices-part-2-why-is-this-happening/

RET. (2018), REC Prices. Renew. Energy Traders Aust. Retrieved October 11, 2018 from http://retaustralia.com.au/rec-prices/

Reposit Power. (2017), *Battery trial to help Bruny power long weekend.* Retrieved August 5, 2017, from https://www.repositpower.com/media-release/bruny-battery-trial/

Roy Morgan Research. (2017), *Solar electric panels hot items in Queensland and South Australia.* Retrieved from https://www.roymorgan.com/findings/7262-solar-energy-electric-panels-march-2017-201707061419

Saddler, H. (2017), *How consumers got burned on electricity prices: It started with networks: RenewEconomy.* Retrieved July 23, 2018, from https://reneweconomy.com.au/consumers-got-burned-electricity-prices-started-networks-48000/

Simshauser, P., Laochumnanvanit, K. (2011), The price-suppression domino effect and the political economy of regulating retail electricity prices in a rising cost environment (No. 20).

Solar Choice. (2017), *Solar system prices based on live data*. Retreived March 18, 2018, from https://www.solarchoice.net.au/blog/solar-power-system-price

Vorrath, S. (2017), *SA rooftop solar installs surge after statewide blackout: Renew Economy*. Retreived March 9, 2017, from http://reneweconomy.com.au/sa-rooftop-solar-installs-surge-statewide-blackout-87015/?utm_source=RE+Daily+Newsletter&utm_campaign=7781616c7d-EMAIL_CAMPAIGN_2017_03_09&utm_medium=email&utm_term=0_46a1943223-7781616c7d-40390089

2.3 China: bureaucratic and market hurdles to move from a central towards a decentral energy system

Antony Froggatt, Liao Maolin, Wei Shen, and Zhou Weiduo

2.3.1 Introduction

In its submission to the Paris Climate Summit in 2015, China committed to peak levels of CO2 by 2030. Following the ratification of the Paris Agreement in September 2016, the National Development and Reform Commission (NRDC) released the Energy Revolution Strategy (2016–30) to put in place domestic measures to ensure the pledge was met. This included a commitment to lower the carbon intensity of GDP by 60–65 per cent below 2005 levels, increase the share of non-fossil energy of the total primary energy supply to 20 per cent, and increase its forest stock volume by 4.5 billion cubic metres, compared to 2005 levels, by 2030 (Carbon Action Tracker 2018).

Meeting these commitments will be challenging as China's energy system is dominated by fossil fuels, as can be seen in Figure 9, where coal accounted for 60 per cent of energy consumption in 2017. These figures also highlight the rapid increase in growth of energy consumption during the first decade of this century, however it is important to note that this trend has tailed off in recent years. China's National Bureau of Statistics' 2017 assessment of National Economic and Social Development (NBSC 2018) show that the Chinese economy grew by 6.9 per cent during that year, an increase of 0.2 per cent over the previous year. While in 2017, total energy consumption increased by 2.9 per cent, leading to a decrease in energy intensity of 3.7 per cent, carbon intensity fell by 5.1 per cent as a result of a gradual move away from coal. During 2017,

How to cite this book chapter:
Froggatt, A., Maolin, L., Shen, W. and Weiduo, Z. 2020. China: bureaucratic and market hurdles to move from a central towards a decentral energy system. In: Burger, C., Froggatt, A., Mitchell, C. and Weinmann, J. (eds.) *Decentralised Energy — a Global Game Changer.* Pp. 47–62. London: Ubiquity Press. DOI: https://doi.org/10.5334/bcf.d.

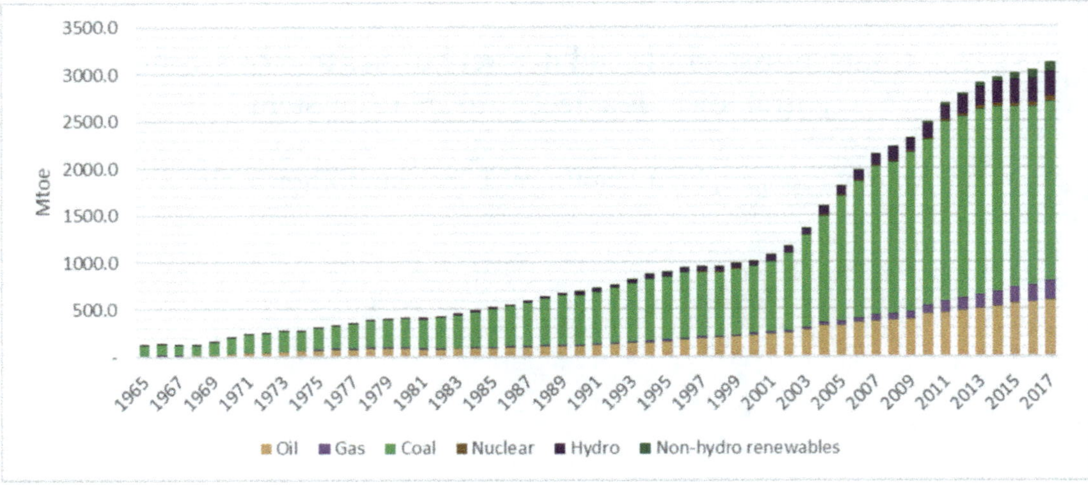

Figure 9: Energy consumption in China 1965–2017.
Source: BP Statistical Review of World Energy (2018).

the consumption of coal grew by 0.4 per cent, which although only slight, was an increase for the first time in three years. There has been an increase in the use of 'non-coal and oil sources' (the government statistics include natural gas, nuclear, hydro, and other renewable sources) which rose to 20.8 per cent of the energy mix, an increase of 1.3 per cent.

There are a number of drivers of change in the energy sector, which have impacts over different timescales. An overarching consideration is the economic transition towards innovation and a more service-led economy. While decarbonisation, along with energy security, are important drivers of change in the energy sector, urban air pollution remains highly influential. (McMullen-Laird et al. 2015) Even the official government statistics suggest that of the monitored 338 cities at prefecture level and above, 29.3 per cent reached the required air standard, however, 70.7 per cent failed to do so. (NBSC 2018) In some cities, such as Beijing, pollution levels have been reduced as a result of the switch from coal to gas and the closure of older factories (Bloomberg 2018).

The grid system is run by two companies, State Grid and China Southern Grid, which have a monopoly of retailing to all customers except large consumers that can self-generate. They are vertically integrated, owning transmission and distribution networks as well as generation and retail. In 2014 the companies were not evenly matched in relation to geographical coverage, size or by assets; State Grid is four times larger than China Southern Grid (Pollitt, Yang & Chen 2017).

2.3.2 Growth in renewable energy

Despite its heavy dependency on fossil fuels and its massive levels of coal consumption China has become a – if not *the* – key country in the global shift

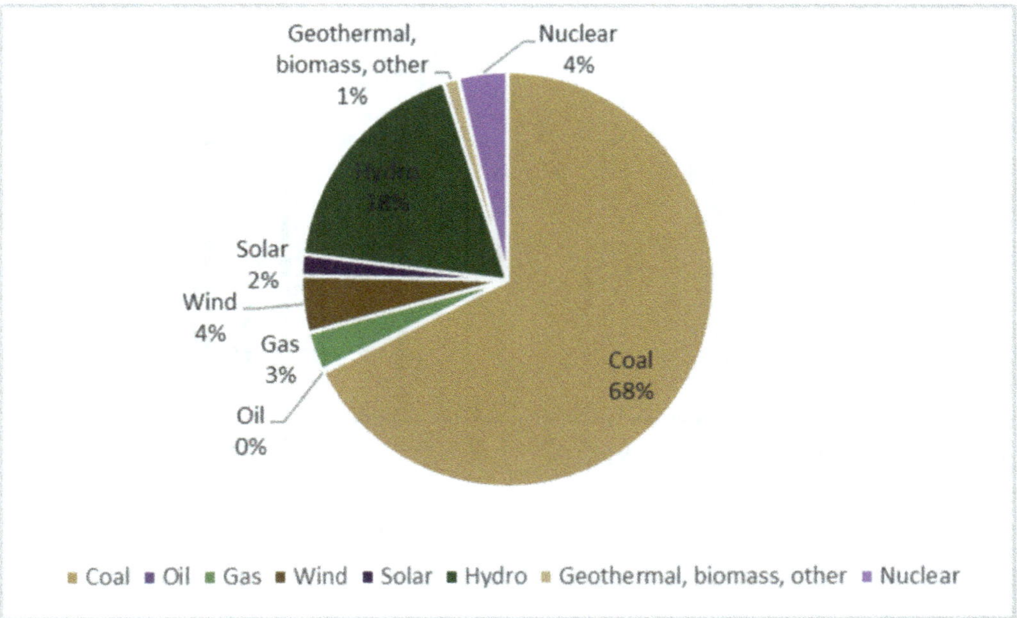

Figure 10: Source of electricity in China in 2017.
Source: BP Statistical Review of World Energy (2018).

towards the greater use of renewable energy, due to its manufacturing base and, more recently, renewable energy deployment rates. China is the world's largest producer of solar PV panels and wind turbines. The annual utilised amount of renewable energy resources in China has steadily increased since 2005. In the power sector in 2017, renewable energy produced 1627 TWh of electricity (25 per cent of the total), with large hydro the largest source, making up 18 per cent of the total, as can be seen in Figure 10.

Hydropower, despite its controversy, remains a significant part of China's energy mix and its growth according to the International Hydropower Association has been 'remarkable'. Although the rate of growth has slowed, in 2017 9.12 GW of capacity was added, leading to a total of 341 GW. Growth for hydro is still on the Government's agenda as new projects, including a 16 GW project in Beihetan, are being developed as it strives to meet the 13th Five-Year Plan target of 380 GW installed capacity by 2020 (IHA 2018).

However, despite the relatively low percentage of power that is coming from new renewable energy, particularly solar and wind, its growth has been rapid and unprecedented globally (see also Figures 11 and 10). In 2017, China added 52.8 GW of solar PV capacity and 19.5 GW of wind, as can be seen in Figure 11.

However, there are concerns in China and internationally about the effect of government measures to control the rate of new solar deployment. In the first half of 2018 the government announced new measures to slow the approval for new subsidised utility-scale PV projects in 2018 and put a 10 GW annual cap on distributed generation (DG) (Renewables Now 2018). The government

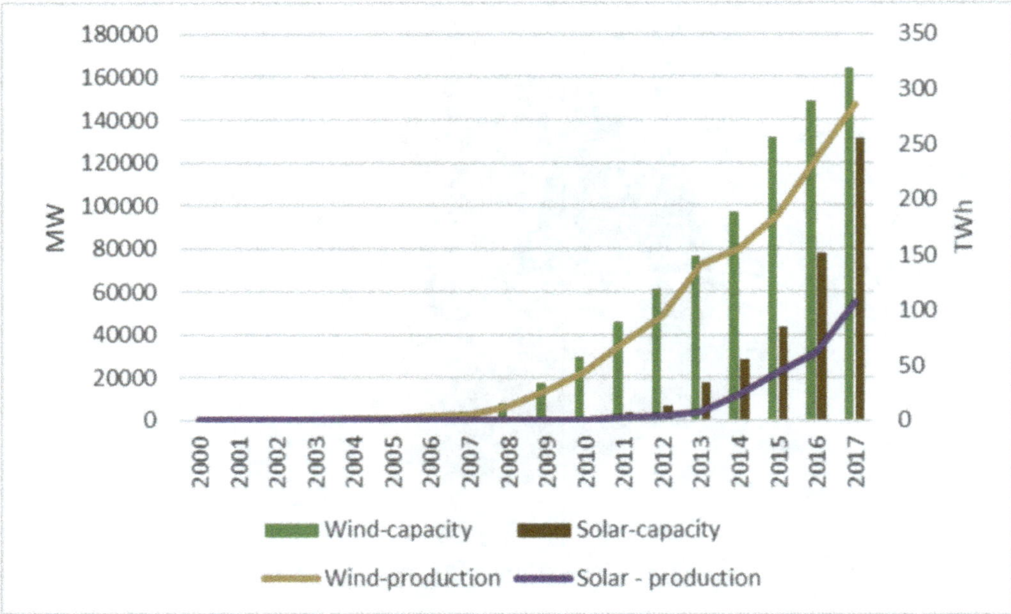

Figure 11: Growth in solar and wind deployment in China (2000–2017)–MW/TWh.

Source: BP Statistical Review of World Energy (2018).

has also cut feed-in tariffs (FiTs) for solar PV and announced that new utility-scale projects would have to compete in auctions.

Furthermore, in mid-June 2018, government announced that biomass and waste to energy plants would no longer be eligible for energy subsidies. (Reuters 2018). Biomass remains an important provider of renewable energy and at present the annual biomass resource for energy use in China displaces about 460 million tons of standard coal (320 mtoe).

2.3.3 Phases of the development of distributed energy development

Over the last decade the rate of connection of renewable energy to the electricity distribution grid in China has fallen behind that in other countries, despite various government attempts to stimulate its growth. During the period from 1990 to 2000, the implementation of distributed energy in various fields and industries was attempted. In China the phrase 'distributed energy', tends to be related to 'combined heat and power generation' or 'combined cooling-heating-power cogeneration', without the specific objective of developing what is now defined as 'distributed energy' more widely in a global sense.

From 2000, some larger distributed energy projects began to be put in place in cities such as Beijing, Shanghai, Guangzhou, etc., however, these were still focused on the development of distributed natural gas. Larger cities were chosen to pilot developments due to the high set up costs, which these areas, with better developed economies and a higher capacity to afford energy prices, were better able to absorb.

The 11th Five Year Plan (FYP) 2006–10 further recognised the need to strengthen the distribution networks for both gas and electricity, and for the further development of combined heat and power and 'distributed cogeneration' with clean energy as the fuel will be developed. Furthermore, energy storage and integrated technologies for heat, power, and refrigeration for micro gas turbines were defined a key frontier technology for the FYP. The plan recognised the potential of renewable energy, including wind, biomass, and solar and the need to scale up to increase commercialisation. Renewable energy was also to be developed for rural energy development, with specific targets on the development of solar thermal, small-scale wind, and biogas digesters. (NDRC 2007)

The 12 FYP (2011–15) called for the further development of large and decentralised renewable energy. This included medium and small hydro resources, including pump storage, as well as solar, biomass, and geothermal. It also called for the strengthening of the grid and the effective development of wind. Specific installation targets were proposed for wind (70 GW) and for solar (5 GW), which were met and exceeded, with installed wind capacity in 2015 at 131 GW and solar at 43 GW. The plan also calls for the government to 'promote the extended application for distributed energy systems' (NDRC 2011).

In 2013, the State Council issued 'Several Opinions on Promoting Healthy Development of the Photovoltaic Industry' (hereinafter called Development Opinions), which proposed to vigorously explore the distributed photovoltaic power generation market. This encourages all types of power users to construct distributed photovoltaic power generation systems in accordance with Power Line Communication. Furthermore, it not only has clear requirements to promote the integration construction of photovoltaic building and large-scale demonstration and application of distributed photovoltaic power, but also encourages the promotion of photovoltaic power in the areas of urban street lighting, urban landscape, communication base stations, traffic lights, etc.

Distributed generation and in particular solar is also seen as a tool for poverty alleviation. In October 2014, the Solar Energy for Poverty Alleviation Programme (SEPAP) jointly issued by the National Energy Administration and the State Council suggested that carrying out photovoltaic poverty alleviation projects, by utilising barren hills and agricultural greenhouses in poverty-stricken areas or facilities to construct photovoltaic power plant, could directly increase the income of the poor. In April 2016, the National Energy Administration along with five further departments, jointly issued Opinions on the Implementation of Photovoltaic Power Generation to Alleviating Poverty to further clarify the specific rules and regulations about the implementation of photovoltaic poverty alleviation projects.

The current and 13th FYP 2015–20 for Energy Development, published by the National Energy Administration, based on the larger Economic and Social

Development plan, further develops the concept of decentralised energy and sets more ambitious targets for the deployment of renewable energy. The plan set proposals for total energy consumption from DG to grow by more than 2.5 per cent per year and for the energy intensity of the economy to improve by 15 per cent. Non-fossil energy consumption should reach more than 15 per cent; natural gas consumption should reach 10 per cent, and the proportion of coal consumption should fall to below 58 per cent. During this period there should be 210 GW of wind power, and 110 GW of solar, of which 60 GW, should come from distributed sources. By the end of 2017, installed wind capacity had reached 164 GW, so was on track to meet its target, while solar had already exceeded it, with 131 GW.

The 13 FYP also calls for clean energy technology development, including the support of smart grid, energy micro-grid, electric vehicle and energy storage technology, and the development of distributed energy networks including the acceleration of the development of the smart grid, and the active promotion of the intelligent substation, and intelligent dispatch system construction (NEA 2016). According to the Renewable Energy 13th FYP, by 2020, total RE electricity installations will reach 680 GW, with electricity production of 1900 TWh (in 2017 this had reached 1627 TWh) and account for 27 per cent of electricity production. The plan proposes that by 2020, the wind tariff should reach grid parity, meaning that the feed-in price for wind will be the same as for coal plants. In parallel to the main Energy FYP, there are 14 supporting FYPs, such as the Renewable Energy 13th FYP, Wind FYP, Electricity FYP etc., which were all released around the same time (Livzeniece 2017).

2.3.4 Lessons from the five-year plans

The IEA and others note that the distributed energy goals, particularly those for gas, have not been met, which is 'unusual in China'. The IEA highlight that to accelerate the deployment of distributed energy a number of improvements could be made which include (IEA 2017):

- Setting out detailed development goals for distributed energy at different government levels: in particular, for national government to provide guidance to local provincial and municipal governments for subnational plans to harmonise with the national plans.
- Distributed energy integration policies: while there may be a recognition of the need to develop distributed generation and that it needs to be connected to the grid, there remain disputes on how and who pays for the connection. Therefore, 'the development of distributed energy will be greatly promoted by introducing standards on integration and operation to clarify the technical demand, procedures for integration and the obligations and responsibilities of each stakeholder'. This can be achieved in China through defining

the obligation of the grid companies to provide access and integration services, to simplify integration procedures, and to strengthen regulation.
- Coordinating the development of energy infrastructure: optimising decentralised energy needs to ensure that the scale and layout of new sources needs to be matched both the renewable resource and the existing distribution network.

2.3.5 Developing distributed solar energy systems in China: challenges and prospects

The Chinese solar PV industry has gone through several developmental stages since the late 1990s (Zhang, Andrews-Speed & Ji 2014).

Solar PV was initially encouraged for lighting and other off-grid activities to enable energy access for remote and nomadic communities. In 1996 the national government introduced a programme designed to help those communities without access to the grid in Western China, through the Brightness Programme. This was expanded between 2002–2007 to North-Western China, through the Renewable Energy Development Project, which reached an estimated two million people. In 2009 the government further promoted the use of solar with the Golden Sun Programme, to encourage its use in the Tibetan plateau (Geall, Shen & Zeren 2017).

However, these developments were relatively small, and before 2012 Chinese solar panels were often sold for export, particularly to the EU and the US markets. Yet the EU and the US trade dispute with China has had devastating impacts on Chinese panel exporters (Meckling & Hughes 2018). In 2013, the Chinese government launched ambitious feed-in tariffs for developing domestic solar energy market (State Council 2013) to rescue the Chinese solar panel manufacturers in crises amid rising protectionist policies (Lewis 2014; Shen 2017). China has since become the world's largest solar nation. Since 2015, China has moved ahead of Germany, becoming the largest investor in solar energy capacities with a total of 131 GW installed in 2017. However, the majority of these investments are for large-scale solar parks, and the distributed solar system[5] only accounts for 13 per cent of the total installed, with around 10.32GW (NEA 2017). This is in contrast to other big solar nations like Germany and the United States, where distributed systems take the lion's share of the market. An interesting question being, why is there such a notable difference?

---

[5] In the Chinese context, distributed solar systems refer locally consumed small-scale projects with 35 KV or below voltage level for grid connection, which includes both roof-top systems and mini solar parks. The electricity can be used either for self-consumption or for sale.

Currently, there is a dramatic (yet diminishing) gap between the manufacturing capacity of PV panels and their deployment into power generation facilities such as solar parks. In 2016 Chinese companies produced 53 GW solar PV modules and panels (Ministry of Industry and Information Technology 2017), while annual deployment within China was around 35 GW (NEA 2017). The 18 GW gap is mainly for exports, particularly to the emerging markets and, much less nowadays, to the western markets. Considering there are many Chinese manufacturers still not operating at their full capacity, the actual gap between manufacturing and domestic deployment capacity could be even higher. As a result, there has been a constant pressure of capacity over-supply in Chinese solar industries throughout the years of its development. Such pressures drove PV panel producers to focus on the large-scale projects to increase sales in the short run. In addition, for project developers and investors, large projects are also preferred due to their economies of scale. Small and distributed systems are often associated with high transaction costs. Lastly, most of China's North Western provinces, such as Xinjiang, Qinghai, Ningxia, and Gansu, are vast and sparsely populated, with sufficient sunlight intensity, therefore ideal for developing large solar parks.

The preference for large-scale projects over small-scale distributed solar systems is a consequence of a combination of market rules and regulations in China, as this is a consequence of geography and pressure on manufacturers to install a lot of panels very quickly. In particular, policies and regulations at the provincial and local level can impact upon the rate of deployment, especially relating to housing regulation and grid access, which can slow down the deployment of renewables at the distributed level. It has become clear, however, that focusing only on large-scale solar parks in the barren Western and remote inland regions is not sustainable in the long-run. The mushrooming solar parks in these areas present tremendous pressure for the grid companies to accommodate all the newly installed solar capacities, as some of them are far from the main grid networks.[6] Even if some of these can be connected, electricity cannot be consumed within the local region as Western provinces are relatively under-developed and the energy demand is consequently low. Yet, long distance transmission to the coastal developed provinces is not only economically unviable but also technologically difficult. As a result, curtailment of energy generation from solar parks has been increasingly rampant.[7] Therefore, Chinese policy makers believe distributed systems are the future of the industry

[6] Although China achieved 100 per cent rural electrification in 2016 (Geall, Shen & Zeren 2017), the capacity and reliability of the grid connection in many places is still inadequate to meet the accommodation and distribution requirement of large renewable energy facilities.

[7] In some provinces, the curtailment can be as high as 30 per cent, which means one third of installed capacity would remain idle and not generating any electricity.

and have been designing various policy tools since 2015 to promote this.[8] The key purpose of these new policies is to 'distort' market preferences for large projects over distributed solar system.

Some concrete measures for promoting distributed solar systems include:

- Provinces with existing curtailment problems must be prohibited from construction of large-scale projects; distributed systems only to be allowed.
- Simplification of the approval process for distributed systems to lower the bureaucratic hurdles. Government approvals on land use and access to rooftops, and environmental or social impacts are to be exempted.
- Grid companies to be required to provide unconditional grid connecting services to distributed solar systems.
- Additional subsidies or tariff support to be provided by both national and local government. In China, there have been significant delays in the payment of subsidies. However, the subsidy payment for distributed solar systems is prioritised.
- Access to upfront capital.
- Some localities legally require that high energy consumption enterprises have to install roof-top solar systems; some also require that all the newly constructed rooftops, once exceeding a threshold area, are to be installed with solar systems; generated 'clean energy' from rooftop solar installations would be counted towards the contribution made by Corporations for achieving energy saving and emission reduction targets as set by the government annually.
- Distributed solar systems to be further promoted as a poverty alleviation programme.[9]

These policies only take effect gradually. In 2016, the frenzy of investment in large-scale infrastructure in the north-western provinces was restricted, with their share of newly added capacity dropping to less than 30 per cent. Meanwhile, there has been a notable increase of investment in distributed systems, mainly among the eastern coastal provinces. The annual instalment of distributed solar capacities doubled compared to the previous year, reaching a

[8] Since late 2014, NEA issued several specific policies to promote and regulate distributed solar systems (NEA 2016; NEA 2017). These policies require local government to provide more generous subsidies and assured grid connections. NEA senior officers have stated clearly in different public speeches and media conferences regarding their preference of distributed system over large-scale solar parks.

[9] In 2014, China announced an ambitious plan to help alleviate rural poverty through the deployment of distributed solar systems in poor areas. The solar energy for poverty alleviation programme (SEPAP) initiative aims to add over 10 GW capacity and benefit more than 2 million households from around 35,000 villages across the country by 2020.

record high of 4 GW. However, there are still significant technical, political, and economic constraints for the further development of distributed systems, as explained in the following section.

Roofs

It is estimated that only 30 per cent of buildings in urban areas have suitable roofs. In rural areas the situation is even worse, which has become a major problem for companies trying to implement SEPAP (solar energy for poverty alleviation) programmes.

In addition to issues of quality, the allocation of ownership, distributing revenues of electricity sales, and sharing the payment obligation are all reasons why rooftop solar installation is not as high as it might be; these are barriers that must be addressed.

Grid connection

As mentioned, rural grids are often less robust in China, and this significantly constrains the expansion of distributed solar systems at village level. In addition, local grid officers also often lack sufficient knowledge and expertise to deal with growing applications for household solar systems, and consequently are reluctant to accept applications for connecting the distributed system. Delaying or denying applications is also not unusual.

Subsidy payment

The dramatic expansion of China's renewable energy industries, particularly in the wind and solar energy sectors, has put the renewable energy subsidy system on the brink of collapse.[10] China's renewable energy subsidy is paid from a national fund that collects additional charges from energy end users. Yet the explosive growth of the solar and wind markets far exceeds the fund's revenue, which has created a mounting shortage of the subsidy. The unofficial estimation of the deficiency of the subsidy can be around RMB 50 to 60 billion by 2016 (Xinhua News 2017). The delay of subsidy payments to the project developers can be several years. In addition, Chinese regulators wish to reach grid parity for wind and solar energy by 2020. Therefore, significant reduction in the subsidy is expected in the coming years and is starting to be seen despite the strong lobbying efforts made by the industry. This expectation is likely to significantly affect investor decisions in the years to come.

Project finance

As distributed projects became increasingly popular, innovative financial arrangements emerged, such as a leasing contract, carbon credit finance, and

[10] Currently, the subsidy programme, Renewable Energy Fund, is mainly collected from the industrial electricity bills.

the energy performance contract (EPC). However, access to project finance is still the largest barrier for many investors in distributed solar systems. Unlike the wind energy sector, where the investors are mainly giant state-owned utility corporations, the investors in solar energy facilities (particularly the distributed systems) are small or medium private enterprises (SMEs). In China's unique political economic system, it is usually difficult for these companies to get access to state-controlled banking services. In addition, the transaction costs and repayment risks attached to distributed systems are very high; projects are dispersed and small in size compared to other energy infrastructure and have very long project cycles for regaining the initial investment, often beyond 15 or even 20 years.

The policy implications for distributed solar systems in China are:

- **First**, clearer policies and regulations are needed to solve the technical issues of the quality of building roofs in rural areas, plus ownership any other legal issues associated with urban residential buildings.
- **Second**, integrating more diversified incentive mechanisms to encourage financial institutions to be actively involved in supporting the distributed systems. Climate, development and energy policy tools, such as carbon offsets, poverty alleviation funds, or green energy certificates, may help to enhance the financial prospect of the distributed system even if government subsidy is to be reduced. However, the integration of these policy instruments requires tremendous coordination among various segments of bureaucratic systems, which is always a big problem in China due to its rather fragmented political system (Andrews-Speed 2012; Mertha 2009).
- **Thirdly**, the grid strategy requires adjustment to present a friendlier approach for distributed systems, prioritising grid upgrade for local consumption and distribution. If all these policy changes were realised – although it is still unrealistic to expect distributed solar systems to dominate the market development in the near future – they would surely play a more important role in China's fast changing energy landscape.

2.3.6 Creating system flexibility

A widely accepted narrative around China's solar and wind development is that the introduction of clear and binding targets in Europe for the deployment of renewable energy and the subsequent rapid deployment of new technologies, led to the establishment of large manufacturing capabilities within China. Then as prices fell and the market slowed in Europe with the economic downturn in 2008, China adopted, and has exceeded domestic targets to maintain its manufacturing capabilities. (Gang 2015). This strategy has led to the global domination of both manufacturing and deployment capabilities.

China is now looking to dominate the manufacturing of the other 'new' technologies to enable the integration of renewables including smaller decentralised generation. In particular, the focus is on batteries, both static and mobile, and electric vehicles.

In 2018, global sales of electric vehicles topped one and a half million for the first time, of which around 1 million were in China. The majority of these sales were from domestic manufacturers, with only 25,000 of these sales from imported vehicles (EV Volumes 2018). As EV sales are expected to expand to 11 million in 2025 and 30 million in 2030, so the need for batteries will increase, leading to lower prices (BNEF 2018), which will benefit grid level storage. In addition, the introduction of smart charging and potential two-way flows of power to EVs will aid system flexibility. China is also dominant in the manufacture of batteries, producing three quarters of the global capacity.

According to Benchmark Mineral Intelligence, China's dominance in the manufacture of batteries is set to continue with 26 large-scale manufacturing plants under-construction and due to be in operation by 2021. These have a combined capability of producing 344.5 GWh of batteries per year (compared to today's capability of 100 GWh), half of which is being built in China (BMI 2017). Whilst a number of the world's largest EV manufacturers have announced they are opening up manufacturing bases in China, including Tesla, BMW, and Volkswagen.

Stationary storage has not, to date, been given the same level of policy support, with a tendency towards other balancing mechanisms such as high voltage transmission (IEA 2016). However, there are examples of significant piloting of new developments, including the construction of 3 MW/12 MWh vanadium redox flow battery (VRB) in Zaoyang, Hubei Province, which is expected to be the test for a 40 MWH project which in turn will be superseded by a 500 MWh project (Energy Storage 2017). There are also further lithium-ion storage projects under development across China, including a 6 MW project by BYD and a 3 MW from China Aviation.

China is also the world leader in the deployment of smart meters with upwards of 469 million in the autumn of 2017; over two thirds of the global total. In 2016 China also exported 130 million units. (Smart Energy International 2018).

2.3.7 Conclusion and prospects

The energy sector in China is changing although, due to the large number of coal power plants, even unprecedently large deployment of new renewable and other technologies result in slow changes in the supply mix. However, there has been a gradual acceptance of distributed renewable sources, especially photovoltaic power. The pattern of the distributed energy industry is maturing, promoting the transition of the power grid in China in the direction of intelligence and micro-grids, as well as boosting poverty alleviation development in China.

The current structure of the power sector and administrative control restricts the development of renewable energy in general, including distributed energy. The reform of the energy market in China is long overdue. Although the reform of the energy industry in China has taken the first step with liberalisation of the coal market, oligopoly competition for oil and gas, and operational (although not ownership) separation of generation and grid, the market-oriented reform is far from complete.

A major barrier to the development of DER remains the lack of a pricing mechanism that accurately prices its value to the system. Therefore, renewable deployment relies on either direct subsidies or fixed prices to ensure development. As with other countries, the Chinese government is keen to reduce the financial support given to renewables as they move towards parity with other generators. This is both understandable and expected. However, it is important that this move is undertaken transparently and in such a way that does not lead to the abrupt halt of the industry. This would have unwelcomed knock-on implications for manufacturers and installers. In the case of China this could also have global implications. The cuts in support for solar PV in the first half of 2018 may well have a disruptive impact across the solar sector in China, and possibility internationally.

China's electricity grids are controlled by two monopolies – the State Grid and China Southern Power Grid – and although policies have been introduced to enable direct power purchases for large users, there are still significant tariff and non-tariff barriers for connections for smaller-scale and distributed generation. This is partly due to a lack of clarity over the roles of the grid operators.

China has many of the components to put decentralised energy front and centre of its energy policy. Furthermore, with large manufacturing capabilities for renewables, storage and smart systems, China also has the capability to capitalise on global trends. However, without governance reforms, enabling smaller actors to enter the market, through regulatory reform and further price disclosure, the opportunity to capture this massive global market and avoid wasted investment could be lost.

2.3.8 References

Andrews-Speed, P. (2012), *The governance of energy in China: transition to a low-carbon economy*. Palgrave Macmillan.

Bloomberg. (2018), *China Is winning its war on air pollution, at least in Beijing*. Retrieved July 16, 2018, from https://www.bloomberg.com/news/articles/2018-01-11/china-is-winning-its-war-on-air-pollution-at-least-in-beijing

BMI. (2017), *Rise of the Lithium ion battery mega factories: What does 2018 hold?* Retrieved July 17, 2018, from see http://www.benchmarkminerals.com/where-is-new-lithium-ion-battery-capacity-located/

BNEF. (2018), *Electric vehicle outlook 2018, Bloomberg New Energy Finance, June 2018*. Retrieved from https://about.bnef.com/electric-vehicle-outlook/#toc-download

BP. (2018), *Statistical review of world energy*. Retrieved from https://www.bp.com/en/global/corporate/energy-economics/statistical-review-of-world-energy.html

Carbon Action Tracker. (2018), *China, climate action tracker*. Retrieved March 13, 2018, from http://climateactiontracker.org/countries/china.html

Energy Storage News. (2017), *Chinese government's strategic push for energy storage to yield large flow battery projects*. Retreived July 17, 2018, from https://www.energy-storage.news/news/chinese-governments-strategic-push-for-energy-storage-to-yield-large-flow-b

EV Volumes. (2018), *China plug-in sales for 2017-Q4 and full year – update – 2018*. Retrieved July 17, 2018, from http://www.ev-volumes.com/country/china/

Gang, C. (2015), China's solar PV manufacturing and subsidies from the perspective of state capitalism. *Copenhagen Journal of Asian Studies*. Retrieved from https://rauli.cbs.dk/index.php/cjas/article/viewFile/4813/5239

Geall, S., Shen, W. & Zeren, G. (2017), *Solar PV and poverty alleviation in China: rhetoric and reality*. STEPS Centre Working Paper no. 93, Brighton: STEPS Centre.

IEA. (2016), *Tracking clean energy progress 2017, International Energy Agency*. Retrieved from https://www.iea.org/publications/freepublications/publication/TrackingCleanEnergyProgress2017.pdf

IEA. (2017), Prospects for distributed energy systems in China. Paris: International Energy Agency.

IHA. (2018), *2018 Hydropower status report, 24 May 2018, International Hydropower Association*. Retrieved from https://www.hydropower.org/publications/2018-hydropower-status-report

Lewis, J.I. (2014), The rise of renewable energy protectionism: emerging trade conflicts and implications for low carbon development. *Global Environmental Politics, 14*(4), 10–35.

Livzeniece, L. (2017), *China's new five-year energy plan, Global Wind Energy Council*. Retrieved July 17, 2018, from http://gwec.net/chinas-new-five-year-energy-plan/

Meckling, J. & Hughes, L. (2018), Protecting solar: global supply chains and business power. *New Political Economy, 23* (1), 88–104.

Mertha, A. (2009), "Fragmented authoritarianism 2.0": political pluralization in the Chinese policy process. *The China Quarterly, 200*, 995–1012.

Ministry of Industry and Information Technology (MIIT). (2017), *The operational situation of solar PV industry (in Chinese)*. Retrieved from http://www.miit.gov.cn/n1146290/n1146402/n1146455/c5505841/content.html

McMullen-Laird, L., Zhao, X., Gong, M. & McMullen, S.J. (2015), Air pollution governance as a driver of recent climate policies in China. *Carbon and Climate Law Review, 9*, 243–255.

NBSC. (2018), *Statistical communiqué of the People's Republic of China on the 2017 National Economic and Social Development, National Bureau of Statistics of China*. Retrieved from http://www.stats.gov.cn/english/PressRelease/201802/t20180228_1585666.html

NDRC. (2007), *11th five year plan on energy development, National Development And Reform Commission*. Retrieved from https://china.usc.edu/national-development-and-reform-commission-percentE2-percent80-percent9C11th-five-year-plan-energy-development-percentE2-percent80-percent9D-april-2007

NDRC. (2011), *12th five-year plan (2011-2015) for National Economic and Social Development, National Development And Reform Commission, translation by the British Chamber of Commerce in China*. Retrieved July 17, 2018, from https://policy.asiapacificenergy.org/node/37

NEA. (2016), *13th five-year plan for energy development, National Energy Administration, translation by Asia and Pacific Energy Forum*. Retrieved July 17, 2018, from https://policy.asiapacificenergy.org/node/2918

NEA. (2017). *National Energy Administration, statistics of solar energy development in 2016 (in Chinese)*. Retrieved from http://www.nea.gov.cn/2017-02/04/c_136030860.htm

Pollitt, M., Yang, C.-H. & Chen, H. (2017), *Reforming the Chinese electricity supply sector: lessons from international experience, Energy Policy Research Group, University of Cambridge*. Retrieved from https://www.eprg.group.cam.ac.uk/wp-content/uploads/2017/03/1704-Text.pdf

Renewables Now. (2018), *GTM cuts by 40 percent solar forecast for China*. Retrieved July 17, 2018, from https://renewablesnow.com/news/gtm-cuts-by-40-2018-solar-forecast-for-china-615667/

Reuters. (2018), *China cuts subsidies for some renewable power projects: finance ministry*. Retrieved July 17, 2018, from https://www.reuters.com/article/us-china-renewables-subsidy/china-cuts-subsidies-for-some-renewable-power-projects-finance-ministry-idUSKBN1JB09X

Shen, W. (2017), Who drives China's renewable energy policies? Understanding the role of industrial corporations. *Environmental Development, 21*, 87–97.

Smart Energy International. (2018), *China leads smart electric meter market*. Retrieved July 17, 2018, from https://www.metering.com/industry-sectors/smart-energy/china-smart-electric-meter-market/

State Council. (2013), *Opinions on deepening the institutional reforms on income distribution*. State Council Policy No. 2016(6). Retrieved from http://www.gov.cn/zwgk/2013-02/05/content_2327531.htm (in Chinese)

Xinhua News. (2017), *The 13th five year plan to tackle the difficult challenge of renewable energy subsidy and curtailments.* Retrieved from http://news.xinhuanet.com/energy/2017-07/31/c_1121404901.htm (in Chinese)

Zhang, S., Andrews-Speed, P. & Ji, M. (2014), The erratic path of the low-carbon transition in China: Evolution of solar PV policy. *Energy Policy, 67,* 903–912.

2.4 Denmark: centralised versus decentralised renewable energy systems

Frede Hvelplund and Søren Djørup

2.4.1 Introduction

In 2012 the Danish parliament voted by a large majority for Denmark to become 100 per cent renewable by 2050 (Klima- Energi og Bygningsministeriet 2012). This decision was the result of a conflict laden political energy transition process (Hvelplund 2013), which had been underway since the first oil crisis in 1973. It was, at the normative level, a Danish 'end of the beginning' of a transition from the fossil fuel era to an energy system based upon renewable energy technologies. The 2012 decision has been confirmed – again with a large majority – by the Danish parliament in an energy agreement dated 20th June 2018.

However, good intentions and political aims are not enough to meet this target. For the transition from fossil fuels to a system based on 100 per cent renewable energy and energy conservation to succeed, there must be a better understanding of the needed technical and institutional changes and its regulatory framework (Hvelplund & Sperling 2018).

The political process to deliver both the renewable energy capacity already in existence by 2012 and the later commitment to 100 per cent renewable energy generation did not flow smoothly. For development to proceed in the period 2018–2050, a number of areas must be addressed, including: development of new policies, the competition and conflicts between interest groups, and acceptance questions regarding new renewable energy projects.

How to cite this book chapter:
Hvelplund, F. and Djørup, S. 2020. Denmark: centralised versus decentralised renewable energy systems. In: Burger, C., Froggatt, A., Mitchell, C. and Weinmann, J. (eds.) *Decentralised Energy — a Global Game Changer.* Pp. 63–81. London: Ubiquity Press. DOI: https://doi.org/10.5334/bcf.e.

This ambitious goal for 100 per cent renewable energy by 2050, can only be reached with continued policy support for the development of both energy conservation measures to reduce consumption and the implementation and integration of renewable energy technologies that deliver a fluctuating supply (Hvelplund, Østergaard & Meyer 2017; Hvelplund & Djørup 2017; Mathiesen et al. 2015). Policies to deliver new renewable technologies, such as wind turbines, solar energy, and biomass energy (Ridjan et al. 2013), are, on their own, inadequate to enable the successful integration of an intermittent supply into the energy mix. However, despite recent governmental activity demonstrating a greater recognition and acceptance of such requirements inherent in this shift to renewables (Energinet.dk 2015), this has not been followed up with an accompanying shift to concrete policies to support the necessary inter-sector integration that is required (Hvelplund, Østergaard & Meyer 2017). In their place there is an unclear assumption that the increasing amounts of renewable energy sources can be managed by building additional electricity interconnectors from Denmark to Holland, the United Kingdom, Germany, etc. At the same time, the integration of the supply side and energy conservation measures have not yet been sufficiently dealt with (Energinet.dk 2016).

For the following discussion it is worthwhile realising that the paradigmatic character of the ongoing change is a transition from relatively scarce greenhouse gas emitting stored fossil fuels that can be used when needed, to clean and abundant renewable energy sources that are fluctuating, and have to be harvested, when the sun shines and the wind blows, and stored so that they can be used when needed. It is worth noting that the switch between these models of energy generation can be made more easily with the support of a small amount of biomass-based energy production (Connolly et al. 2013; Lund et al. 2011).

This country report focuses on the organisational consequences of such a transition from fossil fuels to mainly fluctuating renewable energy sources and increased energy conservation, and considers questions such as: can this change be managed within a centralised model where surplus wind power is exported to neighbouring countries through a network of power interconnectors, where wind power plants are mainly owned by the large former fossil fuel power companies; alternatively, should the transition rely on a decentralised model with smart energy systems and flexible energy consumption delivered by integrating heat, power, transportation, biomass, and energy conservation; furthermore, should this be organised by cooperatively owned wind power plants synchronising supply-side and energy conservation investments? These questions help shape the agenda of current development of the Danish energy system.

Before discussing the merits of decentralisation versus centralisation, we will analyse what can be learned from the energy transition that has already taken place from 1975 to the present, dividing this analysis into two phases from 1975 to 2000, and 2000 through to the present.

2.4.2 *Phase 1: 1975 to 2000 – the development of efficient single renewable energy technologies*

From the mid-1970s to around 2000, the main focus was on the development of cost efficient and well-functioning single renewable energy technologies such as biogas plants, wind turbines, solar heating technologies, etc. In this period, renewable energy only had a minor share of both heat and power production (Hvelplund 2013). The result was that wind power only produced around 3–5 per cent of total electricity consumption annually in the mid-1990s, and in windy periods, not more than 10 to 15 per cent. Wind power was a minor player on the field of energy generation, meaning that fossil fuel plants could continue with business as usual in the existing fossil fuel based infrastructure.

This first phase of development of modern renewable technologies began 45–60 years ago, with roots back to wind powered electricity in 1903 (Thorndahl 2009) and Poul La Cours experiments with hydrogen storage of energy by electrolysis in 1895 (Quistgaard 2009). Today, both photovoltaic and wind power can be produced at a similar cost per kWh to fossil fuel based electricity when external costs are excluded. In fact it is actually much cheaper to produce when external costs are included (Ea analyse 2014).

The development of the first phase is characterised by unstable policy developments with constant conflicts between a centralised model of development based upon the interests of the large fossil fuel-based companies and Danish Industry,[11] and a decentralised model of renewable energy development driven by NGOs,[12] skilled innovators developing new technologies, and small industrial companies (Kooij et al. 2018). It should be noted that this early developmental phase, in the 1970s and beginning of the 1980s, took place against the status quo of a centralised energy policy favoured by the majority of politicians and in the face of strong lobbying by the large power companies (Jensen 2003) and the Association of Danish Industries. These organisations regarded wind power and renewable energy as unrealistic and too expensive, opting instead for coal and nuclear based power production (Beuse et al. 2000; Christensen 1985).

However, in the latter part of this first phase, the share of wind powered electricity production grew to 13 per cent of total electricity consumption and wind turbines were made 'ready' for large-scale deployment with relatively large wind turbines of 2–3 MW. At the same time as these developments were taking place, Denmark saw the creation of around 400 flexible combined heat and power (CHP) systems based on biomass and natural gas, which due to their flexibility in producing energy, provided a more stable environment for integrating wind power, with its intermittent generation levels.

---

[11] The organisation of Danish industries.

[12] The organisation OOA, against nuclear power and OVE, For renewable energy, NOAH, an environmental organisation.

The development and implementation of technically and economically efficient single technologies did not happen automatically, but as a result of 20 to 30 years of technology developments since the mid-1970s, based on a combination of active, small and medium-sized industries, energy focused NGOs, and a democratic process that enabled the policy suggestions from these NGOs to be taken seriously by parliament and implemented despite resistance from the established fossil fuel based power companies (Kruse 1983; Faurby 1982; Hvelplund 1984; Beuse 2000).

One of the most important policies from the beginning of the 1980s was a type of[13] feed-in payment system for wind power sold to the public grid. Another very important policy was that shares in wind power should be local and distributed to many owners; a tax exemption was put in place for incomes from wind power production of less than 150 per cent of the owner's annual electricity consumption. Wind turbine owners were required to live within a 9 km radius from the turbine. In this phase wind power was locally owned (Gorroño-Albizu, Sperling & Djørup 2019), and in the mid-1980s there were between 120,000 and 140,000 local wind power shareholders.

Consequently, wind power gained a strong and widespread political base resulting in parliamentary support despite opposition from fossil fuel companies in the critical period 1987 to 1990, where a 40–80 MW/year home market was required for wind to survive. This was despite the post-1987 collapse of the, roughly 200 MW/year California market, and a situation with almost no world-market sales (Madsen 1988; Beuse et al. 2000).

In this period, Danish wind power survived on a fragile home market due to a continuation of parliamentary support and subsidies for wind power which may not have prevailed without the policy pressure exerted by energy NGOs and the more than 120,000 wind turbine shareholders.

In parallel with this development of wind power that took place up to the early 1990s was an ongoing debate regarding the introduction of small, mainly natural gas-based, CHP units to be established in existing consumer owned district heating systems in small cities. After a longer political dispute, and with resistance from the established power sector, rules were introduced that made these CHP units economical; approximately 400 units were built between 1990 and 1995 amounting to 1.8 GW or 25 per cent of total thermal power capacity. Over the same period, large coal fired power plants lost around 30 per cent of their market share.

What can we conclude from this first phase?

One thing to conclude is that transition takes time, especially when there is no real consensus regarding the direction of development. It took around 40 to

[13] To qualify, this was a price equal to the payment per kWh for a 20,000 kWh/year consumer in their respective Distribution System Operator region.

50 years – from 1974 to 2018 – to develop the current new generation of wind turbines. This development had its roots in the Danish wind power experiences dating back to around 1900 (Christensen 2013). Taking a longer-term perspective, one could say that this current generation of electricity generating wind turbines took 120 years to be realised, with a long fossil fuel period during which wind power development was paused, taking place from 1920 to 1967 or 1977.

From about 1980, when wind power again made its presence felt, development has been slow, but despite resistance from the fossil fuel based power companies, it has been a success. Possibly as a result of its low share of power production, wind power has been viewed as a relatively harmless technology by the members of a hostile fossil fuel-based power infrastructure.

Again, it is important to note that from this first phase of development, wind power and energy conservation in Denmark did not happen automatically but was, to a large extent, driven and developed by small industries and energy NGOs in a conflict-laden political process, often with persistent resistance from the large, established fossil fuel companies, and the Danish Association of Industries (Sønderriis 1998). It is important to emphasise that this development was the result of concrete policies implemented by the Danish parliament in liaison with the relevant actors in society that were, to a large extent, supported by organisations such as energy NGOs, with no vested interests in the old fossil fuel technologies, or in a process that could be described as an 'innovative democratic process' (Hvelplund 2013).

Denmark has a long and strong tradition for non-profit consumer and municipality ownership. This is also the case within the electricity infrastructure which, until 2004, was 100 per cent owned by municipalities and consumers. In 2004 the power plants were sold and subjected to market competition on the Scandinavian Nordpool market. The power plants were sold to the Swedish state-owned company, Vattenfall, and the Danish state-owned company DONG, and thus changed from consumer and municipality ownership to state ownership.

The natural monopoly distribution system operators (DSO) remained consumer and municipality-owned, and subject to a non-profit regulation regime, where energy companies generally do not have the right to use profits for purposes other than to lower consumer prices, although they do have the right to charge consumers for production costs. This could be called a double regulation that, by means of a combination of a non-profit regulation and consumer ownership, provides an incentive for low prices.

This regulation has lately been changed to a new type of double regulation where the non-profit regulation has been replaced by a cost ceiling benchmarking regulation (Hvelplund 2018).

As a result of popular allegiance to consumer and municipality ownership, proposed development of new energy plants in Denmark, especially wind power plants, engender consumer opposition if they are not consumer and/or locally owned. On the other hand, the association of Danish energy companies (Dansk Energy[14]) – a strong lobbying group – has members that include the large power companies Vattenfall and Ørsted (formerly DONG), which have a strong inclination towards Vattenfall and Ørsted ownership of the new electricity production technologies, especially wind power. This has created a situation of conflict within Denmark where the large power companies wish to own the new power production technologies, whilst consumers and those living close to wind power plants tend to accept new wind turbines only if they get a significant share in ownership (Warren & McFadyen 2010).

It is important to emphasise that the consumer ownership share of the value-added chain historically did not include fossil fuel extraction and transportation nor the value-added linked to production of power plants, power grids, etc. In reality the consumer ownership share of power production value added was only linked to the conversion part of a fossil fuel system and amounts to around 25 per cent of the total value-added chain (Hvelplund 2001). It is worth noting that the share of consumer ownership may potentially increase in future renewable based systems, if the fossil fuel share is replaced by a smart energy system consisting of renewable energy supply systems in combination with technologies for the integration of large shares of fluctuating energy.

Finally, it is important to be aware that a district heating infrastructure, covering around 60 per cent of the heat demand, has been successfully developed and implemented. This represents an important part of an infrastructure that potentially integrates both large amounts of fluctuating renewable energy and can embody high percentages of heat conservation and low temperature heat in district heating systems.

2.4.3 Phase 2: the need for an integration infrastructure

It is relatively easy to develop and implement new renewable energy technologies with intermittent supply if they only supply a small share of the energy demand and consumption, because a minor supply from these single technologies can be fitted into the existing energy infrastructure without fundamental changes in the socio-technical energy system.

It is more difficult both at the normative, cognitive, and regulative level to establish energy systems that can handle large shares of fluctuating renewable energy supply. As such, in the second phase of renewable energy development, it is necessary to deal with the development of an integrative energy system that can handle large amounts of intermittent renewable energy supply by means of, amongst other things, integration of power production and heat in district heating systems with heat pumps, solar heating, geothermal heating, etc.,

[14] An association made up mainly of electricity companies.

combined with heat storage systems. Such systems were already discussed 40 years ago, when the need for integrating heat and electricity was analysed and described (Illum 1982). In Denmark, the basic district heating infrastructure to facilitate this integration has, without having the future integration as its purpose, been developed since the 1930s. Furthermore, a renewed expansion of the district heating infrastructure was established at the start of the 1990s both in the large cities and as a part of the 400 decentralised cogeneration plants established between 1989 and 1995. This resulted in an increase of district heating by 40 per cent from 1990 to 2015 (Energistyrelsen 2016). However, there has not been a systematic integration of heat and electricity; heat storage and heat pumps have not been added to the system for example, mainly due to the very high taxes placed on the use of electricity for heat production.

Denmark is now in a phase where integrating heat and electricity is a must, as wind power is becoming the dominant supplier of electricity, necessitating policies for the establishment of an infrastructure that can handle the intermittency inherent within wind power generation. In this phase the question is not any longer, whether a single wind turbine, biomass plant, or photovoltaic unit can produce electricity cheaper than a fossil fuel plant. This is an often used, but wrong comparison. Denmark is in a situation where a 50 to 80 per cent share of energy is supplied by intermittent renewable sources that cannot be compared to single energy supply technologies. Instead, the comparison is between different energy systems that can supply energy when needed, in the right amounts and quantities. This means that we must compare the economics of fossil fuel energy systems with renewable energy-based energy systems, which has been done by Mathiesen et al (2015) and Lund (2014 amongst others. Their work shows that a renewable energy system can be cheaper than a fossil fuel-based energy system, and can deliver energy in the right amounts, at the right time and in the right quality.

But are we establishing the needed infrastructure for the integration of the fluctuating wind power?

In 2015 Danish wind power produced 42 per cent of total electricity consumed, and more electricity in 600 hours of the year than the Danish total electricity consumption during these hours. This is expected to increase to 1400 hours per year in 2020, and already, wind power is exported to countries around Denmark at continuously reduced prices (Hvelplund & Djørup 2017; Bach 2017). At the same time wind powers close to zero short term marginal costs, outcompetes the CHP power production on the Nordpool market, and thus reduces the full-time production hours at both the large and the many small CHP systems, undermining the economic case for these types of production. This is a serious issue, as small CHP systems are likely to become an important part of a flexible infrastructure by supplying electricity in periods where there is only a little wind. Furthermore, the cogeneration plants are to some extent being replaced by inflexible biomass-based district heating systems in the small and large cities of Aarhus and Copenhagen for example, with CHP systems that generate a rather inflexible power supply. Thus, there is an ongoing development towards biomass-based district heating that limits the potential of integrating heat and wind power by reducing the size of the heat market that can be supplied with the less stable supply available from wind power in combination with heat storage and heat pumps. This is caused by an almost zero taxation on biomass for heat, and a high taxation, around €0.05 per kWh, on wind power for heat.

This development regrettably is occurring at the same time as the need to integrate the increasing shares of wind power in a socially and economically efficient way is being highlighted. This is due to the merit order effect (Hvelplund, Möller & Sperling 2013) that has resulted in continuously reduced wind power prices on the Nordpool market (Sorknæs, Djørup, Lund & Thellufsen 2019). Prices went down from around €0.04 per kWh to around €0.03per kWh in the period 2005 to 2015 (Hvelplund & Djørup 2017). This process of reduced wind power prices may continue in the future with plans for a 50 per cent to 60 per cent share of wind powered electricity consumption which could undermine the economics of wind power to an extent that, unless the market is reconstructed, will hamper further wind power expansion (Djørup, Thellufsen & Sorknæs 2018).

The hitherto reason behind district heating has been its fossil fuel efficiency caused by co-production of heat and electricity in cogeneration plants. This reason will be excluded in the future, as a large majority in the parliament in 2012 decided to phase out fossil fuels from the heat and electricity production.

In tandem with this fossil fuel phase out, a set of new argument for a renewable energy-based district heating is developing.

As hot water storage is cheaper by a factor of 100 per MWh for large heating systems compared to electric battery storage systems (Lund et al. 2016), hot water storage for heat at this stage of development are a first step towards handling an increasing share of wind power (with its intermittent supply), for use within the heat market. Hot water storage systems cost approximately €24,000/MWh stored for single houses, and between €500 and €2,500/MWh stored in the larger repositories in a city with district heating, it is a cheaper by a factor of 10–50, to store intermittent energy in district heating systems than in single house systems (Lund et al. 2016).

In cities it is therefore in 100 per cent renewable energy systems more economical to have district, instead of single house, heating systems. Also, for district heating systems, it pays to have low temperature systems because of the increased efficiency in the heat pumps and the reduced loss in the district heating network. Therefore, the system for integrating fluctuating wind power in Denmark is a low temperature district heating system in combination with heat pumps, and hot water storage systems. Furthermore, it should be noted that district heat pipes of good quality have a technical lifetime of around 50 to 70 years.

Along with this development on the energy supply side, it is important to note that in the envisioned 100 per cent renewable energy system of 2050, heat conservation of around 40 per cent compared to current heat use per m2 appears economically optimal (Lund et al. 2016). In transition from a fossil fuel system to a renewable energy system it is becoming technically and economically increasingly important to synchronise supply and demand. Firstly, because Denmark is currently in a transition process characterised by active investors on the supply side and much less active investors in energy conservation on the demand side: this is a serious problem that must be resolved. Secondly, because heat conservation must be implemented 'in time' to avoid costly over-investments in the new renewable energy-based supply system for an uninsulated heat market. Thirdly, because the temperature in district heating systems should be lowered to 50-60 degrees celcius, as low temperature heat supply increases the efficiency of heat pumps (ie. the Coefficient of Performance (COP)), and of solar heating, geothermal heating, and low temperature heat from industries. This reduction in temperature can be implemented without having to invest in larger district heating pipes, if the heat consumption is reduced by heat conservation. This means that Denmark is now in an era where the synchronisation of investments in heat supply systems and energy conservation is increasing in importance.

What can we conclude from the second phase?

The rationale or justification for district heating is changing from being based on *energy efficiency in a fossil fuel-based cogeneration system* to being based on technology *that can handle both an increased share of fluctuating renewable energy*, and the implementation and use of a variety of renewable energy-based fuels (Lund et al. 2014).

It is important not to forget that this ability on the supply side should be underpinned by a systematic technical and organisational synchronisation of investments in the heat supply and heat conservation sides. Heat conservation should also further both low temperature systems that increase the efficiency of heat pumps and the use of industrial waste heat and be implemented in time to avoid overinvestments in supply side systems. This establishment of a new rational base for district heating may also help wind power from an ongoing steady fall in electricity prices resulting from the merit order effect (Hvelplund, Möller & Sperling 2013).

The first step in creating a rational economic basis for wind power is to increase its market size by integrating electricity and heat and thereby enabling wind to enter the heat market. Later steps should be taken to integrate electric based transportation and establish wind to fuel systems (Ridjan et al. 2016;Lund & Kempton 2008; Mathiesen et al. 2015). In this way, by integrating wind power into a smart energy system, the economics of wind may improve to a level where it escapes falling prices created by the merit order effect, meaning that it may pay to build the needed wind power capacity to deliver the 100 per cent renewables by 2050 scenario. However, this can only happen with a taxation policy where taxes on wind power for heat are set at the same level, or lower, than taxes on the scarce resource, biomass for heat. This is not in place today, where tax on wind electricity is high in comparison to the levy of zero tax on biomass for heat.

2.4.4 Decentralised smart energy systems versus centralised power transmission line scenarios

In this second phase of the transition to renewable energy, we have arrived at a crossroads where we are confronted with a choice between a mainly centralised development with large transmission lines and wind power plants owned by large power companies, or a decentralised smart energy system (Lund 2014) development with integration of heat and electricity, electric cars, etc., supported by local and regional ownership.

What follows, is a discussion of these divergent second phase strategies.

The decentralised integration paradigm: development of a smart energy system

When analysing the centralisation versus decentralisation question, it is useful to make an adequate description of the techno-economic character of a smart energy system.

Fossil fuels are stored energy that can be used when needed, however, the investment and management of the extraction of stored fossil fuel energy is only a possibility for large energy companies, *and therefore is an inherently centralised technology.*

The nature of renewable energy is that it is an intermittent source and must therefore be harvested when available. This necessitates the existence of an integrated infrastructure that can either store the energy for later use or transport it by means of interconnectors to other regions or countries where there is an energy need.

In a decentralised smart energy system (Lund et al. 2012), the storage feature of fossil fuels is replaced by coordination and integration technologies and facilities in smart energy systems.

Instead of a distant fossil fuel supply chain with extraction, transportation, and refining located far away from consumers, a smart energy system is established with investments in district heating systems, heat pumps, solar panels, heat storage, energy conservation, electric cars, wind to fuel systems, etc. This means that a value-added share in coordinated and integrated technologies based closer to consumers, replaces a relatively distant fossil fuel based value-added chain.

Consequently, smart energy system technologies may inherently be more suitable for decentralised socio-technical solutions than the fossil fuel-based system it replaces.

The question then, is what characterises the organisational and economic requirements linked to the development and implementation of smart energy systems?

First, smart energy systems need a multifaceted governance system that furthers investments in, and management of, integration technologies. As such, investments in district heating, heat storage systems (Ridjan et al. 2013), heat pumps, and solar heating should be structured in such a way that they can cope with intermittent renewable energy technologies, and must be combined with 'in time' investments in energy conservation. In later stages, the establishment of infrastructure for electric vehicles, wind to fuel systems, geothermal energy etc., will be needed.

The development and implementation of a smart energy system also requires coordination and collaboration between owners of wind turbines, the TSO (Transmission Supply Operator), district heating companies, power distribution companies, and the municipalities and central legislative authorities. This coordination is much more multifaceted than 'just to' develop cost efficient renewable energy single technologies, and requires new organisational models that can develop, implement, coordinate, and manage these many transaction activities, both with regard to long-term investments and day-to-day management. This presents a complex and potentially difficult task of coordination, possibly from a distance.

Due to proximity to consumers it is reasonable to presuppose that the complex coordination and integration involved in smart energy systems, both at the investment and the operation and management levels, may have lower transaction costs in a decentralised than in a centralised governance model (Hvelplund & Djørup 2019). This hypothesis is supported by both transaction cost theory in the Coaseian tradition (Coase 1937; Coase 1988) and the epistemological arguments against central planning in the Austrian tradition (Hayek 1937; Hayek 1945). As coordination becomes more complex, it becomes increasingly costly to convey the adequate level of information to a distant central planning agency – whether industry or government.

Large companies may find themselves, as a result of the following, hamstrung by relatively high transactions costs in any transition to smart energy system solutions. This is due to, amongst other reasons, the fact that they would have to:

- Buy the local consumer and municipality owned district heating systems, which would be very difficult, as these companies are municipality or consumer owned and governed by a non-profit or consumer profit regime. Consumer profit means that any company surplus must be paid back to the consumers in the form of lower prices.
- Invest in heat pumps and heat storage systems linked to district heating systems owned by municipalities and consumers, or to make sure that these investments are implemented.

- Develop a multitude of coordination activities such as, dimensioning investments in the different technologies so that they supplement each other's, and concurrently establish the right amount of energy conservation 'in time' with a conservation level that supports the right low temperature district heating systems. Activities that all seem much easier to perform when the owners of the smart energy system components are the same heat consumers that should also implement the heat conservation investments.
- Distant ownership of onshore wind power is difficult in Denmark, as local citizens due to a long historical tradition, want influence upon such plants by means of, for example, a large ownership share.
- Handle politically conflict-laden negotiations between distant potential owners like the Swedish power company Vattenfall, paying no local taxes and supplying no profits to local actors. The local inhabitants of such distant owner models tend to experience the noise and visual disadvantages of energy generation without receiving benefits from the projects. One ongoing case is the conflict between the Swedish state owned power company Vattenfall and the local Nørrekær Enge wind power community (Olsen & Christiansen 2016).

Due to complicated regulations and the requirements in terms of communication, large distant power companies may struggle to design appropriate investment schemes and to operate these in an efficient way in accordance with local wishes, capabilities, and technological conditions. These companies do not have ownership or control of the smart energy system technologies, nor the ability to handle large amounts of information, in order to behave in a strategic and tactically efficient way. They therefore are comparatively hindered as actors in a decentralised smart energy system.

Consequently, from a political economy point of view, large energy companies may tend to support other more centralised solutions, where their comparative advantages are stronger. Such centralised scenarios will be discussed in the next section.

The centralised on- and offshore wind and transmission line solution

Established power companies, or other distant owners, appear then to face difficulties in implementing and managing smart energy system integration infrastructure. At the same time, these companies and their associations, Danish Energy for example, systematically advocate for 'solutions' that are within the reach of their members within the electricity sector (Energinet.dk 2014). Such 'solutions' mainly consist of offshore wind power in combination with large power transmission grids, which are seen as a way of geographically 'sending surplus wind power to another place' and receiving needed capacity from other countries as reserve power in periods with too little wind power. In Denmark this is combined with large inflexible biomass cogeneration plants that behave almost in the same way as coal fired power plants in the larger cities of Copenhagen and Aarhus. These types of solutions are supported by the Danish TSO, (Energinet.dk 2016), where priority is given to the development of transmission lines between countries. If there is 'too much wind power' in Denmark, it is exported to Germany, the Netherlands, or Scandinavia. When there is too little wind power compared to consumption, additional capacity is imported from neighbouring countries. This is the model in Denmark today, and it is also the model at present proposed for the future, as supported by the TSO, Energinet.dk. Denmark is therefore building, and planning to build, transmission lines to its neighbouring countries without first examining the possibilities and the economics of establishing a smart energy system with a local and regional cross-sectoral integration infrastructure. In practice local and regional cross-sectoral integration is, in the scenarios of the Danish TSO and of the association of Danish power companies (Danish Energy), a second priority.

At present, a centralised transmission line model is favoured by TSOs at both European and Danish levels (Energinet.dk 2017). As suggested, this has already resulted in building large transmission lines to the Netherlands; the COBRA cable received a subsidy of €85.5 million out of a €700 million investment and plans exist to build the Viking cable to the United Kingdom (Djørup 2016; Energinet.dk 2016) which will also receive EU subsidies. The main solution in Denmark today, is this centralised and transmission line model. However, the Danish TSO has also argued for a 'we do both' model, which would mean supporting both the local and regional integration smart energy system model *and* the transmission line model. Sadly, in reality, the smart energy system model is not supported, and in Energinet.dk background reports it is assumed that only around 5 per cent of the heat market will be integrated in 2020 and 15 per cent in 2035 (Energinet.dk 2016), when in fact three times as much would be possible. Meanwhile this centralised transmission line model does not seem to solve the renewable energy intermittency challenge on a long-term basis. With increasing shares of such intermittent renewable energy, Denmark's neighbours are also increasing their wind power capacities and the Danish wind regime is similar in the North European countries (Bach 2017). So far the cost-benefit analysis made by the TSO, and justifying investments in transmission lines has been questioned as not being transparent and disregarding alternative courses of action (Djørup 2016; Lund et al. 2017; Mathiesen, Lund & Djørup 2018). The above analysis indicates that the solution lies in smart energy systems with more flexible electricity consumption delivered by means of local and regional integration of heat, electricity, transportation, wind to gas, etc. (Lund et al. 2012; Mathiesen et al. 2015; Bach 2017).

Solutions that integrate increasing amounts of intermittent renewable energy in a smart energy system are needed and are already supported by thorough calculations showing that it is possible and cheaper to develop and implement a decentralised model, where surplus wind power is integrated into the heat and transportation market (Lund et al. 2012).

2.4.5 Conclusion and policy recommendations

The above discussion of the first phase of renewable energy development has shown that a technological transition takes time and that it does not happen on its own. A sort of innovative democracy is necessary, where the influence from new NGOs independent of the old fossil fuel-based companies is needed. It also demonstrated that the development of single renewable energy technologies is 'easy' in the way that these could be absorbed into the old fossil fuel infrastructures without major changes to these, due to the very minor share of total electricity supply that these single technologies contribute.

In the second phase of renewable energy development, where wind power produces more than 40 per cent of electricity production, the basic infrastructure has to be changed, to absorb large amounts of intermittent energy. As with the changes in phase one, such changes do not happen unilaterally in the inherited market arrangements. Concrete policies are necessary to bring this change about.

The viability of decentralised smart energy system models has been theoretically well documented for decades and practiced to some extent by a few district heating companies. In comparison, the model supporting large heat pumps and large heat storage systems based on wind power is in a start-up phase with only a few projects, and as such, it has not yet reached a full-scale implementation.

A decentralised smart energy system model seems to be able to solve the problems linked to a transition to intermittent renewable energy sources, but under present policies has not been fully realised. The present high tax on wind power for heat and zero tax on biomass for example, mitigates against such a transition.

On the contrary there still is a strong tendency towards realisation of the centralised offshore transmission line scenario in combination with distant ownership of wind power plants and biomass heat based on imported biomass in the largest cities. Furthermore, this development is still subsidised, and in upcoming years, several billion Euros will be invested in interconnectors, despite there being little evidence that this model will solve the problem of integrating increasing shares of intermittent wind power, either technically or economically. Technically because Denmark's neighbouring countries are expanding their wind power capacity and have similar wind regimes, and economically because it relies upon huge EU subsidies and might not be able to compete with smart energy system solutions on an economically equal playing field.

So, it looks as if Denmark and the European Union are in a situation where billions of Euros will be invested in a centralised model that will prove to be a blind alley, incapable of integrating increasing amounts of intermittent energy, and at the same time hindering the development and implementation of a decentralised model that does have the ability to solve the problem of intermittency. It is therefore necessary, to level the playing field to make it possible to establish the new smart energy system infrastructure.

This could be achieved by introducing the following policies:

- An EU directive that introduces a subsidiarity principle where local integration via smart energy systems has priority and, if needed, investments in transmission lines take second priority.
- The European Union should give the same level of subsidies to a decentralised smart energy system infrastructure, as the present EU subsidy provides to transmission interconnectors.
- Denmark's taxes on wind power for heat should be set to at least the same level as taxes on biomass.
- Renewable energy tenders should be designed so that it is possible for local citizens to participate as a foundation, where the profit goes to both local citizens and to the common good of the region.
- Introduce legislation whereby wind power project managers are compelled to sell at least 51 per cent of a wind power project to local consumers, municipalities and local companies; that this local ownership percentage should be kept during the lifetime of a renewable energy project.
- Allocate 30 per cent of the surplus from a wind power project to a foundation with the deed to use the money for the common good. This could be for environmental purposes and for the development of smart energy system integration technologies. It could also be used to support energy conservation in the region.
- Introduce changes to the role of the Danish TSO, Energinet.dk, so that it is obliged to support integration of intermittent renewable energy in accordance with a subsidiarity principle.
- Introduce transmission tariffs, where the payment is a function of consumer use of the transmission system. Payments are currently charged only for transmission in situations of bottleneck limitations in the transmission line.

The implementation of a decentralised smart energy system-based integration of renewable energy could become a realistic possibility if the above policies were introduced at both the EU, and the national level, in Denmark.

2.4.6 References

Bach, P.F. (2017), Getting rid of wind energy in Europe. Retrieved February 10, 2019, fromhttp://pfbach.dk/firma_pfb/references/pfb_getting_rid_of_wind_energy_2017_07_02.pdf

Beuse, E., Boldt, J., Maegaard, P., Meyer, N.I., Windeleff, J. & Østergaard, I. (2000), Vedvarende energi i Danmark. Copenhagen: OVEs Forlag.

Christensen, B. (1985), Officielt er det umuligt! Kolding Højskoles forlag, Kolding.

Christensen, B. (2013), History of Danish Wind Power. In P. Maegaard, A. Krenz & W. Palz (Eds), *Wind power world rise mod. Wind energy*. Singapore: Pan Stanford Publishing.

Coase, R.H. (1937), The nature of the firm. *Economica 4*, 386–405.

Coase, R.H. (1988), The firm, the market, and the law. Chicago: University of Chicago Press.

Connolly, D., Lund, H., Mathiesen, B.V., Østergaard, P.A., Möller, B., Nielsen, S., I., R., Hvelplund, F., Karl, S., Karnøe, P., Carlson, A., Kwon, P.S. & Bryant, S. (2013), Smart energy systems: holistic and integrated energy systems for the era of 100 percent renewable energy. Aalborg: Aalborg University.

Djørup, S.R. (2016), Fjernvarme i forandring. Aalborg: Aalborg University. Retrieved from https://doi.org/10.5278/vbn.phd.engsci.00137

Djørup, S., Thellufsen J.Z., Sorknæs, P. (2018), The electricity market in a renewable energy system. *Energy*. 162, 148-157. https://doi.org/10.1016/j.energy.2018.07.100

Ea Energy Analysis. (2014), Elproduktionsomkostninger-Samfundsøkonomisk langsigtede marginalomkostninger for udvalgte teknologier. Retrieved February 10, 2019, from https://ens.dk/sites/ens.dk/files/Analyser/ea_energi analyse_elproduktionsomkostninger.pdf

Energinet.dk. (2014), *Electricity interconnections*. Retrieved from Energinet.dk

Energinet.dk. (2015), *Energikoncept 2030-baggrundsrapport*. Retrieved from https://energinet.dk/-/media/CF94250F0EF04F3EBE0D0C473590DF5D.pdf

Energinet.dk. (2016), *Energinets analyseforudsætninger 2016*. Retrieved from https://energinet.dk/Analyse-og-Forskning/Analyseforudsaetninger/Analyseforudsaetninger-2016.

Energinet.dk. (2017), *Strategi 2018-2020 – Energi over grænser*. Retrieved February 10, 2019, from https://energinet.dk/Om-publikationer/Publikationer/Strategi-2017

Energistyrelsen. (2016), *Energistatistik 2015-2016*. Retreived from https://ens.dk/service/statistik-data-noegletal-og-kort/maanedlig-og-aarlig-energistatistik

Faurby, P. (1982), Risinge et lærestykke i elværksdemokrati. *Atomkraft 36*, 8–10.

Gorroño-Albizu, L., Sperling, K. & Djørup, S. R. (2019), The past, present and uncertain future of community energy in Denmark: Critically reviewing and conceptualising citizen ownership. *Energy Research & Social Science*, 57. https://doi.org/10.1016/j.erss.2019.101231

Hayek, F.A. (1945), The Use of Knowledge in Society. *American Economic Review, XXXV, 4*, 519–30.

Hayek, F.A. (1937), Economics and knowledge. *Economica, IV new ser.*, 33–54.

Hvelplund, F. (1984), Hindringer for vedvarende energi, F.Hvelplund, F. Rosager, K.E. Serup Aalborg Universitetsforlag.

Hvelplund, F. (2001), Electricity Reforms, Democracy and Technological change, Department of Development and planning, Aalborg University 2001.

Hvelplund, F. (2013), Innovative democracy, political economy, and the transitioin to renewable energy: a full scale experiment in Denmark 1976-2013. *Aplinkos Tyrimai, Inzinerija ir Vadyba, 66*(4), 5–20.

Hvelplund, F., Möller, B. & Sperling, K. (2013), Local ownership, smart energy systems and better wind power economy. *Energy Strategy Reviews, 1*(3), 164–170.

Hvelplund, F & Djoerup, S. (2017), Multilevel policies for radical transition: governance for a 100 percent renewable energy system. *Environment and Planning C: Politics and Space, 35*(7), 1218–1241.

Hvelplund, F., Østergaard, P.A. & Meyer, N.I. (2017), Incentives and barriers for wind power expansion and system integration in Denmark. *Energy Policy, 107*, 573–584. DOI:https://doi.org/10.1016/j.enpol.2017.05.009

Hvelplund, F. & Sperling, K. (2018), Denmark: renewable energy in the member states of the EU, EU Energy Law Volume III, Second. ed. Deventer: Claeys and Casteels.

Hvelplund, F. (2018), Public regulation and consumer ownership of electricity companies. Report submitted to: Instituto Centroamericano de Administración Pública (ICAP), and Autoridad Reguladora de Los Seviciós Publicos (ARESEP), Costa Rica.

Hvelplund, F. & Djørup, S. (2019). Consumer ownership, natural monopolies and transition to 100% Renewable Energy Systems. *Energy*, 181, 440-449. https://doi.org/10.1016/j.energy.2019.05.058

Illum, Klaus. (1982). Energi, organisation og samfund. Aalborg Universitet.

Jensen, I.K. (2003), Mænd i modvind. Copenhagen: Børsens forlag.

Klima- Energi og Bygningsministeriet. (2012). "Regeringens Energi- Og Klimapolitiske Mål — Og Resultaterne Af Energiaftalen i 2020." 2012.

Kooij, H.-J., Oteman, M., Veenman, S., Sperling, K., Magnusson, D., Palm, J. & Hvelplund, F. (2018), Between grassroots and treetops: community power and institutional dependence in the renewable energy sector in Denmark, Sweden and the Netherlands. *Energy Research and Social Science, 37*, 52–64.

Lund, H et al 2011. (2011). *(CEESA) Coherent Energ and Environmental Systems Analysis.* Edited by H Lund. Department of Development and Planning.

Lund, Henrik, Anders N. Andersen, Poul Alberg Østergaard, Brian Vad Mathiesen, and David Connolly. (2012). "From Electricity Smart Grids to

Smart Energy Systems – A Market Operation Based Approach and Understanding." Energy. https://doi.org/10.1016/j.energy.2012.04.003.

Lund, Henrik. (2014). Renewable Energy Systems: A Smart Energy Systems Approach to the Choice and Modeling of 100% Renewable Energy Systems. Second Edi. Academic Press (Elsevier).

Lund, Henrik, Sven Werner, Robin Wiltshire, Svend Svendsen, Jan Eric Thorsen, Frede Hvelplund, and Brian Vad Mathiesen. (2014). "4th Generation District Heating (4GDH)." Energy 68 (April): 1–11. https://doi.org/10.1016/j.energy.2014.02.089.

Lund, H., Østergaard, P.A., Connolly, D., Ridjan, I., Mathiesen, B.V., Hvelplund, F., Thellufsen, J.Z. & Sorknses, P. (2016), Energy storage and smart energy systems. *International Journal of Sustainable Energy Planning and Management, 11*. Retreived from https://journals.aau.dk/index.php/sepm/article/view/1574

Lund, H., Mathiesen, B.V., Hvelplund, F., Djørup, S. & Madsen, H. (2017), *Professorer advarer mod Viking Link: Fremlæg de hemmelige beregninger og få alternativerne belyst. Altinget.dk, 9 November 2017.* Retreived Febrauary 10, 2019, from https://www.altinget.dk/energi/artikel/161436-kronik-el-ledning-til-england-er-en-dyr-og-sort-investering

Madsen, B.T. (1988), Windfarming in Denmark. *Journal of Wind Engineering and Industrial Aerodynamics, 27.*

Mathiesen, B.V., Lund, H., Connolly, D., Wenzel, H., Østergaard, P.A., Möller, B., Nielsen, S., Ridjan, I., Karnøe, P., Sperling, K. & Hvelplund, F.K. (2015), Smart energy systems for coherent 100 percent renewable energy and transport solutions. *Applied Energy, 145*, 139–154.

Mathiesen, B. V., Lund, H. & Djørup, S.R. (2018), *Viking Link er en risikofyldt investering af broget kulør. Altinget.dk.* Retreived February 10, 2019, from http://vbn.aau.dk/files/275147326/20180508_Viking_Link_Svar.pdf, accessed 10 February 2019

Olsen, E. & Christiansen, H. (2016), Stor opbakning til Vindmølleforening. Nibe Avis, 15 March 2016. Retreived February 10, 2019, from https://nibeavis.dk/norrekaerenge/4039-stor-opbakning-til-vindmolleforening

Quistgaard, T. (2009), The experimental windmills at Askov 1981-1903. In *Wind power-the Danish way.* The Poul La Cour fouondation – Askov 2009, p. 87.

Ridjan, I., Mathiesen, B.V., Connolly, D. & Duić, N. (2013), The feasibility of synthetic fuels in renewable energy systems. *Energy 57*, 76–84.

Sorknæs, P., Djørup, S., Lund, H., Thellufsen, J.Z. (2019), Quantifying the influence of wind power and photovoltaic on future electricity market prices. *Energy Conversion and Management 180* , 312-324.

Sønderriis, E. (1998), *Dansk Industri vil tage energien fra Auken. Information.dk, 8 July 1998.* Retreived February 10, 2019, from https://www.information.dk/1998/07/dansk-industri-tage-energien-auken?vwo_exp_badges=|32|

Thorndahl, J. (2009), Electricity and wind power for the rural areas 1903-1915. In *Wind power - The Danish way*. The Poul La Cour fouondation- Askov 2009, p. 87.

Warren, C.R. & McFadyen, M. (2010), Does community ownership affect public attitudes to wind energy? A case study from south-west Scotland. *Land use policy 27*, 204–213.

2.5 Germany: from feed-in-tariffs to auctions and the question of diverse actors

Dörte Ohlhorst

2.5.1 Summary

The stakeholder structure of the formerly oligopolistic German electricity market has changed significantly during the energy transition process in the last two decades. The big energy supply companies have lost high market shares to smaller, heterogeneous actors that generate renewable electricity. Both the structure of electricity generating technologies and the ownership structure of the facilities are increasingly decentralised. In this chapter about Germany, the term 'decentralised' is used in the context of community-scale renewable electricity generation and ownership.

The increasingly decentralised system has advantages in terms of resilience, social acceptance, democratic participation in the use of universal goods, and the development of value added in municipalities and regions. This chapter describes how decentralised renewable energy in the hands of communities and citizens has developed in Germany. It elaborates on how the political framework has recently changed, which actors have gained influence and which effects can be expected from the changes in the funding regime regarding decentralisation, local and regional development, democratisation of energy business, the variety of actors and stakeholders, as well as regarding the involvement of citizens in renewable energy supply.

2.5.2 Decentralised citizen energy in Germany

Although primary energy consumption is still dominated by fossil fuels (about 79 per cent in 2018) and some nuclear energy (6.4 per cent in 2018), Germany

---

How to cite this book chapter:

Ohlhorst, D. 2020. Germany: from feed-in-tariffs to auctions and the question of diverse actors. In: Burger, C., Froggatt, A., Mitchell, C. and Weinmann, J. (eds.) *Decentralised Energy — a Global Game Changer.* Pp. 82–100. London: Ubiquity Press. DOI: https://doi.org/10.5334/bcf.f.

aims to obtain 60 per cent of its total energy (final energy consumption) and 80 per cent of its electricity generation from renewables by 2050. The government has goals for phasing out nuclear by 2022 and simultaneously decarbonising the economy by reducing greenhouse gas emissions by 80 to 95 per cent of 1990 levels by 2050. This transformation is known as the 'Energiewende' (BMU/BMWI 2011), which contrary to popular opinions, is more than just a reaction to the Fukushima nuclear accident.

The beginnings of what is now known as 'Energiewende' date back decades. The transformation was strongly driven from the bottom up. Social movements starting in the late 1960s, such as the peace movement with regard to nuclear weapons, the anti-nuclear power movement, and the environmental movement triggered a change in awareness in large parts of German society. These movements were closely linked to the events of the oil, nuclear, and environmental crises of the 1970s and 1980s (e.g. oil price crises in 1973 and 1979–1980; nuclear accidents in Three Mile Island in 1979 and Chernobyl in 1986; smog and air pollution, especially in coal mining regions and large cities). They mobilised significant resistance against the prevailing policy positions and economic conventions of the time, and activated social engagement for structural changes in energy policy and the supply system. With the founding of the German Green Party (1980) the topic was carried into the political system. Decades of critical social debates about the existing energy policy, starting in the late 1960s and continuing into the present, have led to a counter-proposal to conventional energy supply and resulted in its transformation. The strong desire of many citizens to be concretely involved in the energy transition can only be understood in light of this history (Schreurs 2008).

In Germany, renewables accounted for less than 12 per cent of electricity consumption in 2006. At the end of 2017, they accounted for 38.5 per cent, a materially important expansion achieved in just a few years. Decentralised renewable energies, supported or initiated by citizens, made a significant contribution to this success. Many citizens in Germany have the desire to be involved in the value creation and employment of the energy transition (EWSA 2015; Dunker & Mono 2013). Financial participation – as an active shareholder or passive donor – provides an opportunity for profitable investments. From the ownership of renewable generation facilities local added value can be generated and jobs as well as apprenticeship training positions can be created. A further component of participation is an active citizen engagement in the sense of not only a financial but also a conceptual initiative to implement renewable energy projects.

In Germany, citizens have organised themselves to initiate and operate renewable energy projects. Currently, almost half of the renewable electricity generation capacity subsidised by the Renewable Energy Sources Act (EEG) (except for offshore wind) is in the hands of small private investors. These actors include individuals, households, energy cooperatives, civil law partnerships, limited liability companies or limited partnerships, decentralised initiatives in municipalities and regions, farmers and civil wind farms. These diverse forms of citizen energy ('Bürgerenergie') are an important driver for the dynamics of the energy transition. Moreover, citizen energy encourages more diverse acceptance and support for renewables and thus increases the stability of the German renewable energy sector (BBEn 2014; Zuber 2014; Debor 2014; Müller & Holstenkamp 2015; Ott & Wieg 2014, see also country report on Denmark).

§ 3 (15) EEG 2017 (Renewable Energy Sources Act 2017) defines a citizen energy company as an entity,

- 'Which consists of at least ten natural persons with voting right,
- in which at least 51 per cent of the voting rights are held by natural persons which live in the urban or rural district in which the onshore wind energy installation is to be erected,
- in which no member or shareholder of the undertaking holds more than 10 per cent of the voting rights of the undertaking' (Renewable Energy Sources Act 2017).

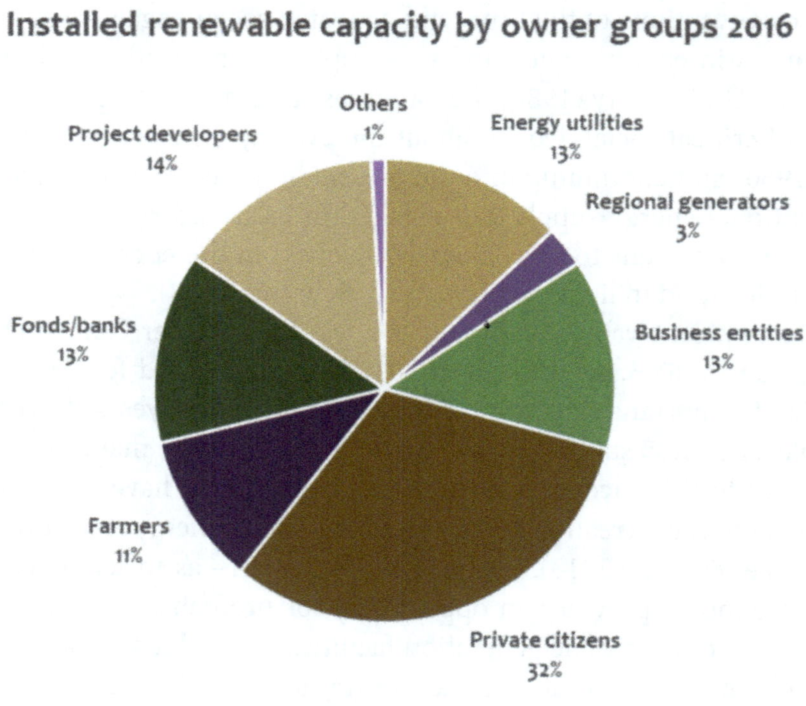

Figure 12: Installed capacity of renewable energy by owner group in Germany in 2016.

Source: Author's contribution based on figures published by trend:research 2018[15]

[15] The distribution of ownership to these groups may change in the future due to the shift from fixed feed-in tariffs to tenders for renewable projects. The number of new start-ups of energy cooperatives has fallen by 60 per cent in 2014 due to uncertainties with the new funding system and because the obligatory direct marketing for generating facilities with 500 kWp or more makes the entry less attractive (Kahla et al. 2017).

Cooperatives and companies in which small private investors hold at least 50 per cent are summarised under the term 'citizen energy', see figure 12, (trend: research & Leuphana University of Lüneburg 2013). Cooperative models are a particular legal form: on the one hand they are an economic actor, and on the other a civil society actor. Energy cooperatives have a democratic legal structure, in which the members are protected from the dominance of majority owners. Each member, regardless of the level of participation, has one vote in the General Assembly and has a say on the use of resources. Members to a large extent come from the locality or region. The deposits are typically long-term investments. Investors have some protection against financial loss, because each member is only liable for their deposits. Secondly, each cooperative must belong to an auditing association, through which the finances of the cooperative are regularly checked. The cooperative business model is usually long term and requires less return on the investment than other business models with high short-term profit goals. The debt ratio is lower, compared to other legal forms, and the control by supervisory board members and members is often more intense. It is true that cooperatives are not safe from bankruptcy, but this happens less frequently to cooperatives than to conventional companies. Profits are often reinvested in the cooperative goals. The interest on cooperative deposits, though, is lower compared to the return to the investors on their deposits in stocks and risky bonds.[16]

Citizen energy is characterised by a high degree of identification with the local energy supply because most of the electricity is produced close to the consumption points in the neighbourhood (Zuber 2014).[17] The commitment and investment of citizens represents a key driving force of the Energiewende: citizen energy has a market share of at least 31.5 per cent of the installed renewable electricity capacity in Germany (see Figure 12; trend:research 2018; Zuber 2014). Between 2007 and 2014 the total number of registered energy cooperatives in Germany grew from 94 to 973 and has since increased more than tenfold. In the years 2011 (195 newly registered cooperatives), 2012 (187), and 2013 (172), the increase in energy cooperatives reached an all time high (Müller & Holstenkamp 2015). In the year 2012, cooperatives invested €1.67 billion in renewables, and built up electricity generation capacities with an output of 933 MW (DGRV 2015; Kemfert & Schäfer 2012; trend: research & Leuphana University Lüneburg 2013). The German energy cooperatives founded since 2006 have about 130,000 members (DGRV 2015; Janzig 2015).[18] Hubert Waiger from

[16] http://www.sozialinvestieren.de/blog/insolvenzsicherste-rechtsform-genossenschaften-bleiben-unangefochten-sieger/

[17] There are also German cooperatives that are committed to the repurchase of local electricity networks or the concession for the operation of local networks. However, these are not directly affected by the amendment to the EEG.

[18] Not all energy cooperatives are also citizen energy cooperatives. The DGRV data and the database from Degenhart, Holstenkamp, and Müller are not directly comparable.

the Federation for Environment and Nature Conservation Germany (BUND) called this commitment the 'Largest civil movement in the history of our country'.[19] Cooperatives are seen as a possible economic model for eco-social transformation and as a learning environment for civil societal and democratic values (Walk 2014).

The high number of renewable energy projects in the hands of citizens, organisations, and affiliated companies has led to a pluralistic stakeholder and ownership structure in the German power generation market. Farmers, public utilities, and small and medium enterprises complete the spectrum of power generators, which has expanded enormously since the expansion of renewables (Moss, Becker & Naumann 2014).

Previously, the electricity market was divided between a few large energy utilities. The German government repeatedly declares the target to maintain actor diversity in energy transition efforts and that regional and local efforts towards a low-carbon energy transition are welcome. This also corresponds to the coalition agreement between the CDU/CSU parliamentary party (Christian Democrats) and the SPD parliamentary party (Social Democrats), which promises that the local communities will be more involved in the added value of renewable energy facilities and the opportunities for project participation of citizens will be improved.

One advantage of raising capital through the private investment of citizens is the lower yields that are deemed acceptable by this group, as opposed to larger companies (Mono 2013). The enhanced decentralised participation in financing can then ensure a large part of the investments necessary for the expansion of renewable energy. However, the mobilisation of private capital required to meet investment needs requires suitable investment models and incentives that need to be set by governance (Jacobs et al. 2014; WBGU 2012).

2.5.3 Drivers for – and against – decentralised renewable energy in Germany

The national funding regime has acted as a central driver in the development of renewable energy in Germany. With the Electricity Feed-In Act and its successor, the Renewable Energy Sources Act (EEG), a priority for grid connection and electricity feed-in as well as fixed feed-in tariffs for renewables was introduced. The Electricity Feed-In Act, passed in 1990, was one of the first laws in Europe that obliged public energy utilities to purchase and remunerate

The former only covers energy cooperatives founded from 2006; the latter recognise registered energy cooperatives.

[19] Public Hearing of the Committee on Environment, Nature Conservation, Building and Nuclear Safety; June 4th, 2014. https://www.bundestag.de/dokumente/textarchiv/2014/eeg-novelle/281434.

renewable electricity on a yearly fixed basis. During the 1990s, not only the renewable electricity support policy, but also the general legal framework for the energy sector was important for the promotion of renewable energy in Germany. The Renewable Energy Sources Act was adopted in 2000, replacing the Electricity Feed-In Act and establishing a new, pioneering support policy with improved investment security for generators: while under the Electricity Feed-In Act compensation rates were expressed as percentages of average customer tariffs, the new rates were now fixed for 20 years. The EEG has been amended regularly since the year 2000. The feed-in tariffs set out in the law have been the central prerequisite for the expansion of renewables during the following 15 years. The renewable energy share in electricity supply increased dynamically – today it covers more than one third of the power supply. Due to the declining, but guaranteed feed-in tariffs, investments in renewables were able to be deployed with low risks for investors.

Besides the funding regime, a strong citizen engagement favoured the production of decentralised renewables in Germany. Citizen engagement is motivated by a desire for autonomy, freedom of action, control, and for shaping one's own living conditions. People have relatively high confidence in the objectives and values of citizen initiatives, whereas there is increasing distrust towards state actors and short-term-profits-oriented economic operators, and the non-transparent relations between them (Büscher & Sumpf 2014; Sumpf 2014). One motivation for self-generated electricity is (partial) energy independence, and for some households it is even the idea of a self-sufficient power supply. A consumer survey in 2013 showed that six per cent of all Germans are so-called prosumers – they do not only consume, but also generate electricity (both for own consumption and for sale, Verbraucherzentrale Bundesverband 2013).

The decentralised initiatives are improving local value added income and employment options (Deutscher Landkreistag 2014a). In 2012, citizen energy projects brought a (gross) value of €5.3 billion to the respective regions (Hauser et al. 2015). Adverse local developments like declining municipal revenues due to shrinking population numbers can be compensated through wind, solar and biogas projects (Deutscher Landkreistag 2014b). About 113,600 permanent jobs, particularly in the field of plant operation, are assigned to citizen renewable energy projects (Hauser et al. 2015).

It is assumed that participation in citizen energy projects increases not only local acceptance of renewable energy plants, but also the experience of self-efficacy as well as social inclusion of vulnerable people (Tews 2018). However, acceptance depends on whether the implementation of the projects correspond with the ideas of citizens about democratic decision-making processes, transparency and participation, sufficient decision options, an inclusive planning culture, and an adequate distribution of costs and benefits (Mono 2013; Bauknecht, Vogel & Funcke 2015; Schweizer et al. 2014; 100 prozent erneuerbar stiftung 2012; Bovet & Lienhoop 2015).

Due to recent changes in the political and legal framework however, decentralised initiatives appear to be at risk. The changes have resulted in the termination of price-based feed-in tariffs and premium systems. Moreover, a shift from an optional to a mandatory direct marketing system for new renewable energy plants has taken place. The obligatory direct marketing presents new challenges for suppliers and operators.

2.5.4 Change of funding regime by amendment of the Renewable Energy Sources Act (Erneuerbare Energien-Gesetz)- EEG in 2014

With an amendment to the EEG in 2014, the government set the course for the introduction of a tendering model that replaces the system of fixed feed-in remuneration (e.g. Kahles 2014). Previously, the level of the feed-in tariff (differentiated by technology and based on forecasts) was determined by parliament, but the actual cost could differ from the forecasts. As a result, the level of financial support could prove to be excessive or too low. According to the German Government, the introduced bidding procedure will address this matter.

In the new procedure, annual quotas of renewable electricity generating capacity are tendered via auctions. First experiences with the tender model where gained in a pilot phase for photovoltaic ground-mounted systems (Deutscher Bundestag 2014). From 2017 onwards, the compensation rates for all renewables are determined via a tender procedure.[20] Only those market participants that have been selected by the tender process are allowed to build renewable energy plants. The aim is to carry out the construction of new renewable energy installations as economically as possible and in accordance with European state aid rules. Small systems with a capacity below one MW, however, are not subject to the tendering procedure.

The change in the funding mechanism was carried out not only due to national but also supranational pressure by the European Commission (Beermann & Tews 2016; Tews 2014; Vogelpohl et al. 2017). The Commission considers quota systems as the most appropriate instrument for an integrated energy market. Its harmonisation efforts were intensified at the end of 2013, as the European Commission revised state aid guidelines and consequently initiated an infringement procedure against the German funding system with fixed feed-in tariffs. Large energy utilities and actors pursuing a market-liberal policy in the European Commission supported this policy.

Not only on the European level, but also in Germany, the control of costs and capacities of renewables has been increasingly debated. Influential players

---

[20] The feed-in tariff, which is determined through the tender procedure, is adapted to the dynamics of renewable energy expansion when the expansion has exceeded or fallen below the corridor of annual quotas during the reference period ('principle of breathing lid').

argued that the promotion of renewable energy should be better adapted to the existing market structure (Wassermann, Reeg & Nienhaus 2015). By adopting and fundamentally changing the funding system, the German federal government (a grand coalition) thus not only pursued the aim of decarbonising the electricity production in compliance with the EU law, but also had the intention to improve the predictability of the expansion of renewable energy dynamics and to achieve the government's renewable energy development objectives in, what the government argued was a more cost-effective way. In contrast, the critics of the EEG amendment argued that these objectives could also have been achieved with the existing funding regime. They are afraid that the expansion of renewable energies in Germany will stagnate and, consequently, the climate protection targets cannot be met. The government, however, argued that the new funding scheme can promote more market and technology competition and at the same time avoid the protection for specific technologies.

2.5.5 Decentralised citizen energy and actor diversity at risk?

It is becoming apparent that new risks are arising from the new procedure that was introduced by the German Bundestag, initiated by the Ministry of Economics. It is feared that with the conversion of a regime with fixed feed-in tariffs to an auction based system the financially strong supra-regional providers will have improved opportunities to increase their market share. The results of the pilot bidding rounds on ground mounted PV (2014–2016) verified the concerns regarding a decline in the diversity of actors and the exclusion of small players (Tews 2018). For decentralised, small projects with local character, the amendment constitutes a much bigger challenge (EnKlip 2015; Kahl, Kahles & Müller 2014; Leuphana Universität of Lüneburg & Nestle 2014; BNetzA 2015). The maintenance of actor diversity is a politically defined goal[21] that is both explicitly mentioned in the coalition agreement and in the EEG 2014 (§ 2 para. 5 sentence 3). A specific focus on citizen-based energy companies is officially explained and justified in §36g EEG 2017: 'In particular, locally anchored citizen energy companies have made a significant contribution to the necessary acceptance of new on-shore wind energy projects. Without this acceptance, the expansion of wind energy cannot be achieved in the planned amount' (Deutscher Bundestag 2016, author's translation). Nevertheless, many scientists and practical actors consider the new funding regime a serious threat to the continued commitment of the small, local and civic actors (Niederberger & Wassermann 2015; Tews 2018). It is feared that competition

[21] See § 2 para 5 sentence 3 EEG 2014: 'When switching to the tender procedure, actor diversity in renewable electricity generation is to be obtained' (translated by author).

decreases and the plurality of actors cannot be obtained in the renewable energy market, because:

Participation in a tendering procedure requires high operating expenses, is time consuming, complex, and costly. In coping with the complex requirements large providers have significant advantages over small ones. The transaction costs incurred can be intercepted more easily by larger companies (Leuphana Universität Lüneburg & Nestle 2014; Kahl, Kahles & Müller 2014; Klessmann 2014). Citizen energy actors usually find it more difficult to spread the risk or protect themselves against the risks through their own private capital. They have a lower credit rating and hardly any opportunities for interim financing. This is demonstrated by international experience with tendering mechanisms (BWE 2014).

There is uncertainty to tenderers whether the contract is awarded in an auction participation. Accordingly, uncertainty for investors about the expected returns increases (Grießhammer & Bergmann 2015; Nestle 2015).[22] The return on capital investment equity for investors decreases because the final price they receive for the renewables is lower due to rising costs and additional risks (Grau, Neuhoff & Tisdale 2015). Even if an actor has been awarded the contract, it may occur that the site-specific costs are higher and revenues are lower than projected. As a consequence, the plant may not be realised. In this case, significant penalties may be accrued (Ecofys 2014; Hauser et al. 2014; Hauser et al. 2015).

In the course of compulsory direct marketing, plant operators have to either bring their electricity to market themselves (this can also be to regional markets), or they can employ external services. This results in costs for marketing, forecasting, and profile service. Furthermore, the operator (electricity supplier) incurs many obligations, including the sale of excess capacity and the purchase of additional electricity in the event of equipment failure or under-production, in order to ensure security of supply for customers. The effort and cost can be problematic, particularly for civil energy projects and energy cooperatives, because they generally are run by volunteers and have fewer opportunities to diversify risk (BEE 2015; Leuphana Universität Lüneburg & Nestle 2014). The obligatory direct marketing increases the financial risk for plant operators. They must compensate for uncertainties in marketing with risk premiums. Moreover, it is assumed that banks require long-term power purchase agreements with economically powerful direct marketers for project finance. As part of the credit assessment of the direct marketers, it is possible that large marketers are favoured.

A new characteristic of the tender procedure is not only the determination of the level of compensation via competing offers, but also the waning influence

[22] Also changes of the Investment Code and concession law have contributed to uncertainty (Kahl et al. 2014).

of the parliament: previously, the actual level of remuneration was set through the feed-in tariffs in the Renewable Energy Sources Act, which was developed and changed by parliament. The tender procedure is performed by the Grid Agency, allowing no influence of parliament on the compensation rate.

Overall, the new funding system is characterised by a lower market openness, lower actor plurality, and lower investment security than the previously applied principle of fixed feed-in tariffs. It is therefore feared that tenders and the obligatory direct marketing represent a massive market entry barrier for citizen energy operators. Under the new system, they are hardly able to compete against larger competitors and hardly able to afford the huge transaction costs for developing legally sound and competitive bids (BEE 2015).

There is some evidence for these concerns: since 2014, the formation of new energy cooperatives has been rapidly declining. According to the German Genossenschafts- und Raiffeisenverband (German Cooperative and Raiffeisen Confederation, DGRV), the number of newly registered energy cooperatives in 2014 declined by 60 compared to the previous year. In 2014, only 54 energy cooperatives were founded, whereas in 2011, 167 such enterprises were launched nationwide, see figure 13 (Müller & Holstenkamp 2015).

This sharp decline in growth illustrates the effects of the amendment to the Renewable Energy Sources Act (Grießhammer & Bergmann 2015). Investment activities of the existing energy co-operatives have dropped significantly.

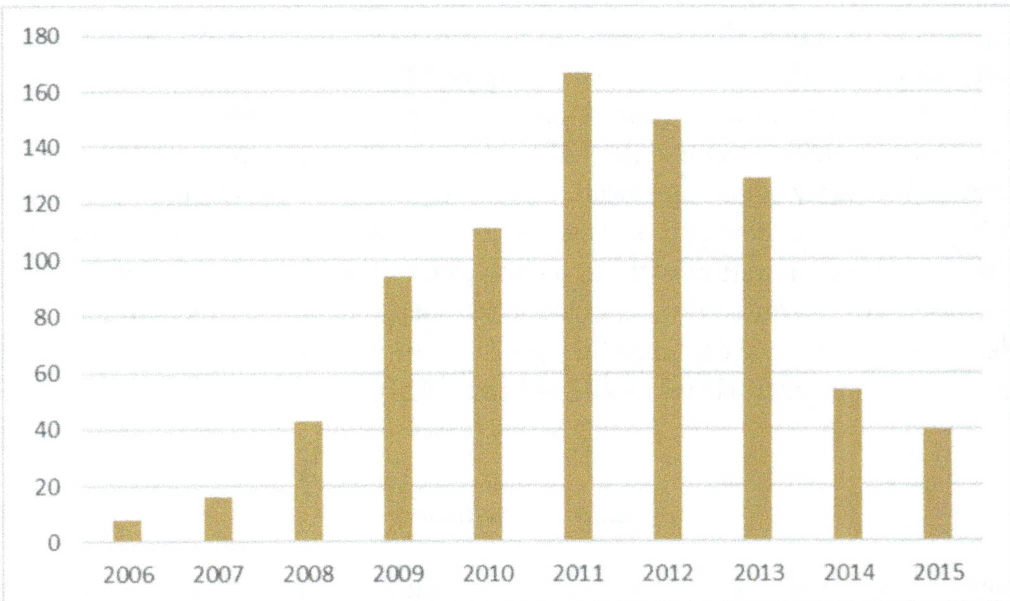

Figure 13: Start-up of energy cooperatives 2006–2015.
Source: DGRV[23]

[23] German Cooperative and Raiffeisen Confederation, DGRV; https://www.erneuerbareenergien.de/buerger-muessen-teil-der-energiewende-bleiben/150/437/88921/.

Surprisingly, in the first auctions in 2017, the majority of the awards were given to citizens' energy companies. The bidders made use of the specific rules for citizen energy: they could take part in the bidding procedure without having a construction permit, and could make use of the longer time span between awards and realisation of the project deadline (up to 4 to 5 years). However, it turned out that the successful bidders, privileged as citizens' energy companies, were set up by a small number of professional project developers, contractors or service providers (Tews 2018). They transformed their projects and found constructions that made them look like citizens' energy projects because they were adapted to the official rules of citizen energy. As such, the project developers could make use of the respective privileges in the bidding procedure. Moreover, they were – at least organisationally – assigned to one single project developer (Bundesnetzagentur 2017). The exceptional rules – intended to support rather unprofessional and small citizen-based projects – were used by business actors in an economically rational way to meet the legal requirements of a citizen energy company. This leads to the need to re-amend the Renewable Energy Sources Act if the goal of preserving the variety of actors is to be achieved and if 'real' citizens' energy projects should be addressed. It also leads to the need of a clear definition of the term 'citizen energy'.

According to the German Renewable Energy Federation (BEE), environmental organisations and green electricity providers, tenders change the level playing field to the detriment of the variety of actors and generate new risks that are difficult to manage. In order to maintain the diversity of actors, the BEE repeatedly demands that projects up to 18 MW should be excluded from the tendering procedure. This approach proposed by the BEE has been confirmed by the EU Competition Commissioner – a derogation to this extent would be permissible under European law (BEE 2016). Until now, only projects up to an installed capacity of one megawatt are excluded from the tendering procedure. It is proposed that new citizen`s energy projects should receive an administratively fixed remuneration without having to go through the tender procedure, because with adequate project sizes, a regional balance between supply and demand and a high acceptance they produce the best power solutions for local and regional markets (Hannen & Enkhardt 2015).

2.5.6 *Conclusions*

Germany has multiple goals for the implementation of the energy turnaround: not only a decarbonisation of the electricity supply by 2050, but also a cost-efficient power supply and a diversity of electricity-producing actors. The fulfilment of the latter goals is important because efficient energy generation at a reasonable price as well as electricity production by community and citizen energy projects helps to gain acceptance for the large-scale transformation with high transition costs and new technologies emerging all over the landscape. As the number of wind farms across the country increases, the acceptance factor becomes increasingly important.

The recent replacement of fixed feed-in tariffs by a technology-specific and volume based auction system, not introduced by a regulator, but by the German Bundestag, represents a fundamental change in the funding regime of the German Renewable Energy Sources Act. The government launched the new funding model (auction system), even though experiences in other countries have shown that auction systems often did not reach the desired results and cost-minimising effects (AEE 2014; Agora 2014; Ecofys 2014; Hauser et al. 2014). It is not clear whether the European guidelines actually require opting out of the feed-in tariff. The legal situation would probably allow wider exemptions for smaller market participants than is provided for in the amended Renewable Energy Sources Act (Münchmeyer, Kahles & Pause 2014). Therefore, it is assumed that the German Government used the EU requirements to realise its own national reform ideas (Vogelpohl et al. 2017). It is expected that the modification of the funding regime will not lead to a return towards centralised power generation technologies, but to a more centralised ownership structure of generating plants.

In the first tender rounds in 2017, almost exclusively citizens' energy companies were awarded. This was a great surprise for all who accompanied the design of the tender scheme in the new Renewable Energy Sources Act. With less than six Euro cents per kWh, the bidding prices where about one cent below the previous average remuneration – and are thus very low. Evidence shows, professional investors have applied under the cover of alleged citizens' firms, who were able to construct their bidding company according to the privileging exception rules for citizen energy.

In the 2019 tender rounds for wind turbines, the level of competition was lower than ever before. A trend that has already been observed over several rounds in 2018 has been continued. In January 2019, with a tendered volume of 700 megawatts (MW), only 72 bids totalling 499 megawatts were submitted. A total of 67 bids with a volume of 476 megawatts were awarded a supplement. Only eleven of these supplements went to citizens' energy companies. The competitive level for the second round of calls in May 2019, with a 55% signing, had a new worrying dimension. Again, insufficient bids were submitted to cover the tendered amount of 650 MW. Of the 650 MW tendered, only 270 MW could be awarded for the construction of new wind turbines. The reasons for this dramatic decline in the expansion of wind energy are manyfold: the complexity of the tendering process, increasing citizen protests against wind energy, a great uncertainty in the market and high penalties in case of failed implementation. The capping of tenders at a level of 2.5 gigawatts and the debate over new regulations regarding the minimum mandatory distances to residential buildings caused uncertainty. In addition, more and more citizens are resisting wind turbines in their neighbourhood – and often take legal action. A project that has been awarded a contract, has a time frame of only two to a maximum of two and a half years to actually realise the wind farm. Otherwise the bidder faces penalties ranging from 150,000 to 200,000 euros per turbine.

It remains unclear whether the future design of the tender rounds actually serves the achievement of the desired wind power capacities on the one hand and the preservation of the diversity of actors on the other – and thus the protection of operating models with local investors, local and regional value added, mixed ownership structure, distributed voting rights, a participatory legal structure, and social inclusiveness. It is possible that the different energy policy sub-processes and the legal framework in Germany will reduce the citizens' commitment to the energy turnaround and lead to more centralised ownership structures of the generating plants. However, there will continue to be citizen engagement for renewable energies in the future, because it is increasingly cost effective to produce electricity with sustainable resources. Therefore, organisational and institutional niches are likely to be found for citizens' cooperatives. In addition, the possibility of self-supply with renewable electricity may increasingly be used in the future. If, however, the detour through the tendering procedure slows down the dynamics of citizen participation, this poses a risk for the acceptance of the energy projects in the electricity sector. A diversity of stakeholders, acceptance, participation and active involvement of citizens are essential success factors for the energy supply in Germany, because these factors help place the energy transition on a broader footing within society.

2.5.7 References

100 prozent erneuerbar stiftung. (2012), *Akzeptanz für Erneuerbare Energien. Akzeptanz planen, Beteiligung gestalten, Legitimität gewinnen (Kurzfassung).* Retrieved February 10, 2019, from http://100-prozent-erneuerbar.de/2012/09/akzeptanz-fur-erneuerbare-energien-%E2%80%93-100-prozent-erneuerbar-stiftung-veroffentlicht-umfangreichen-leitfaden/

AEE Agentur für Erneuerbare Energien. (2014), Studienvergleich: Finanzierungsinstrumente für Strom aus Erneuerbaren Energien. Metaanalyse von Vorschlägen für die zukünftige Finanzierung von Strom aus Erneuerbaren Energien. AEE: Berlin.

Agora Energiewende. (2014), Ausschreibungen für Erneuerbare Energien. Berlin.

Bauknecht, D., Vogel, M. & Funcke, S. (2015), Energiewende — Zentral oder dezentral? Diskussionspapier im Rahmen der Wissenschaftlichen Koordination des BMBF Förderprogramms: Umwelt- und Gesellschaftsverträgliche Transformation des Energiesystems. Ökoinstitut Freiburg, 28.07.2015.

BBEn. (2014), *Energiewende braucht Bürgerenergie (Positionspapier).* Retrieved from http://www.buendnis-buergerenergie.de/fileadmin/user_upload/downloads/Positionspapiere/BBEn_Positionspapier_EEG-Novelle.pdf.

BEE. (Bundesverband Erneuerbare Energie e.V.) (2015), Regierungspläne für Solarausbau bremsen die Energiewende. Pressemitteilung des BEE vom 21.01.2015. http://www.verbaende.com/news.php/Regierungsplaene-fuer-Solarausbau-bremsen-die-Energiewende?m=101121.

BEE. (Bundesverband Erneuerbare Energie e.V.) (2016), *BEE-Bilanz zum EEG 2017. Deutliche Drosselung der Energiewende, leichte Verbesserungen im Detail.* Retrieved February 10, 2019, from https://www.bee-ev.de/home/presse/mitteilungen/detailansicht/bilanz-zum-eeg-2017-deutliche-drosselung-der-energiewende-leichte-verbesserungen-im-detail/

Beermann, J. & Tews, K. (2017), Decentralised laboratories in the German energy transition. Why local renewable energy initiatives must reinvent themselves. *Journal of Cleaner Production (Special Volume: The opportunities and roles of experimentation in addressing climate change), 69*, 125–134.

BNetzA. (2015), Hintergrundpapier. Ergebnisse der zweiten Ausschreibungsrunde für PV-Freiflächenanlagen vom 1. August 2015. Bonn: Bundesnetzagentur.

Bovet, J. & Lienhoop, N. (2015), Trägt die wirtschaftliche Teilhabe an Flächen für die Windkraftnutzung zur Akzeptanz bei? Zum Gesetzesentwurf eines Bürger- und Gemeindebeteiligungsgesetzes in Mecklenburg-Vorpommern unter Berücksichtigung von empirischen Befragungen. *Zeitschrift für Neues Energierecht 16/3*, 227–234.

Büscher, C. & Sumpf, P. (2014, July), *Trust, distrust and confidence in energy system transformation.* Paper presented at XVIII ISA World Congress of Sociology "Facing an unequal world: challenges for global sociology", Yokohama, Japan.

Bundesnetzagentur. (2017), *Pressemitteilung 15.08.2017. Ergebnisse der zweiten Ausschreibung für Wind an Land.* Retrieved fromhttps://www.bundesnetzagentur.de/SharedDocs/Pressemitteilungen/DE/2017/15082017_WindAnLand.html

BWE. (2014), Stellungnahme zum Entwurf eines Gesetzes zur grundlegenden Reform des Erneuerbare-Energien-Gesetzes und zur Änderung weiterer Bestimmungen des Energiewirtschaftsrechts (BT Drs. 18/1304). Berlin: Bundesverband Windenergie.

Debor, S. (2014), The socio-economic power of renewable energy cooperatives in Germany: results of an empirical assessment. Wuppertal Papers, No. 187 (Working Paper). Retrieved from https://www.econstor.eu/dspace/bitstream/10419/97178/1/785254935.pdf

Deutscher Bundestag. (2014), Entwurf eines Gesetzes zur grundlegenden Reform des Erneuerbare-Energien-Gesetzes und zur Änderung weiterer Bestimmungen des Energiewirtschaftsrechts. 18. Wahlperiode. Drucksache 18/1304. Berlin: Deutscher Bundestag.

Deutscher Bundestag. (2015), Auswirkungen der Novelle des Erneuerbare-Energien-Gesetzes 2014. Antwort der Bundesregierung auf die Kleine Anfrage der Abgeordneten Dr Julia Verlinden, Oliver Krischer, Christian

Kühn (Tübingen), weiterer Abgeordneter und der Fraktion BÜNDNIS 90/ DIE GRÜNEN, Drucksache 18/5774. 18. Wahlperiode. Drucksache 18/5898. Berlin: Deutscher Bundestag.

Deutscher Bundestag. (2016): Gesetzentwurf der Fraktionen der CDU/CSU und SPD. Entwurf eines Gesetzes zur Einführung von Ausschreibungen für Strom aus erneuerbaren Energien und zu weiteren Änderungen des Rechts der erneuerbaren Energien. Drucksache 18/8860, 21 June 2016, see http://dip21.bundestag.de/dip21/btd/18/013/1801304.pdf

Deutscher Landkreistag. (2014a), *Energiewende ist für ländlichen Raum. Chance und Herausforderung zugleich — Nachbesserungen beim EEG notwendig.* Retrieved from http://www.landkreistag.de/presseforum/pressemitteilungen/1351-pressemitteilung-vom-19-maerz-2014.html

Deutscher Landkreistag. (2014b), *Regionale Wertschöpfung durch erneuerbare Energien. Handlungsstrategien für Landkreise zur Initiierung einer regionalen Kreislaufwirtschaft. Unter Mitarbeit von Institut für angewandtes Stoffstrommanagement (IfaS).* Retrieved from http://www.landkreistag.de/images/stories/publikationen/bd-120.pdf

DGRV Deutscher Genossenschafts- und Raiffeisenverband e.V. (2015), *Energiegenossenschaften. Ergebnisse der DGRV-Jahresumfrage (zum 31.12.2014).* Retrieved from https://www.dgrv.de/webde.nsf/7d5e59ec98e72442c1256e5200432395/418a5acd4479ba4ec1257e8400272bec/$FILE/DGRV-Jahresumfrage.pdf

Dunker, R. & Mono, R. (2013), *Bürgerbeteiligung und erneuerbare Energien. Kurz-Studie von Beteiligungsprojekten in Deutschland durch die 100 prozent erneuerbar Stiftung. Berlin.* Retrieved from http://100-prozent-erneuerbar.de/wp-content/uploads/2013/07/Buergerbeteiligung-und-Erneuerbare-Energien_100pes.pdf

Ecofys. (2014), Design features of support schemes for renewable electricity. A report compiled within the European project "Cooperation between EU MS under the Renewable Energy Directive and interaction with support schemes". Retrieved from https://ec.europa.eu/energy/sites/ener/files/documents/2014_design_features_of_support_schemes.pdf, accessed 10 February 2019

Engerer, H. (2014), *Energiegenossenschaften in der Energiewende. DIW Roundup – Politik im Fokus. 17. Juli 2014.* Retrieved from http://www.diw.de/de/diw_01.c.470180.de/presse/diw_roundup/energiegenossenschaften_in_der_energiewende.html.

EnKlip. (2015), Ausschreibungen für Erneuerbare Energien — überwindbare Hemmnisse für Bürgerenergie?, see http://www.buendnis-buergerenergie.de/fileadmin/user_upload/Buergerenergie_und_Ausschreibungen_BBEn_2015.pdf.

EWSA Europäischer Wirtschafts- und Sozialausschuss. (2015), *Die Energie von Morgen erfinden – Die Rolle der Zivilgesellschaft bei der Erzeugung*

erneuerbarer Energie. Retrieved from http://www.renewableuk-cymru.com/wp-content/uploads/2015/01/CivilSocietyRenewableEnergy.pdf

Grau, T., Neuhoff, K. & Tisdale, M. (2015), Verpflichtende Direktvermarktung von Windenergie erhöht Finanzierungskosten. DIW Wochenbericht Nr. 21.2015, pp. 503-508. Berlin: Deutsches Institut für Wirtschaftsforschung.

Grießhammer, R. & Bergmann, M. (2015), Wissenschaftliche Koordination des BMBF-Förderprogramms: „Umwelt- und gesellschaftsverträgliche Transformation des Energiesystems. Entwicklungsportfolio, Synthese, Partizipationsmethoden, Transfer. Freiburg: Öko-Institut.

Hannen, P. & Enkhardt, S. (2015, November 26), EEG-Reform: Gemeinsame Ausschreibungen für PV-Dachanlagen und Solarparks geplant. *pv magazine*.

Hauser, E., Weber, A., Zip, A. & Leprich, U. (2014), *Bewertung von Ausschreibungsverfahren als Finanzierungsmodell für Anlagen erneuerbarer Energienutzung. Endbericht. Für IZES GmbH, Saarbrücken*. Retrieved from http://www.bee-ev.de/Publikationen/IZES2014-05-20BEE_EE-Ausschreibungen_Endbericht.pdf

Hauser, E., Hildebrand, J., Dröschel, B., Klann, U., Heib, S. & Grashof, K. (2015), *Nutzeneffekte von Bürgerenergie*. Retrieved from http://blog.greenpeace-energy.de/wp-content/uploads/2015/09/IZES-2015_09_10_B percentC3 percentBCE-Nutzen_Endbericht.pdf.

Hochloff, P., Sandau, F. & Bofinger, S. (2014), *Bürgerenergiewende oder Industrieprojekt? Eine Beleuchtung des Diskurses zur Ausrichtung der Energiewende*. Retrieved from http://www.polsoz.fu-berlin.de/polwiss/forschung/systeme/ffu/forschung-alt/projekte/laufende/11_energytrans/konferenz2014/programm/2-iwes-Hochloff.pdf

Jacobs, D., Peinl, H., Gotchev, B., Schäuble, D., Matschoss, P., Bayer, B., Kahl, H., Kahles, M., Müller, T. & Goldammer, K. (2014), *Ausschreibungen für erneuerbare Energien in Deutschland — Ausgestaltungsoptionen für den Erhalt der Akteursvielfalt*. Retrieved from http://www.iass-potsdam.de/sites/default/files/files/working_paper_ausschreibungen_final.pdf

Janzig, B. (2015), Bürgerprojekte in Deutschland. Politik grätscht dazwischen. Berlin: die tageszeitung (taz), 21.7.2015.

Kahl, H., Kahles, M. & Müller, T. (2014), *Anforderungen an den Erhalt der Akteursvielfalt im EEG bei der Umstellung auf Ausschreibungen. Würzburger Berichte zum Umweltenergierecht. Würzburg*. Retrieved from http://www.stiftung-umweltenergierecht.de/fileadmin/Bilder/Newsletter/WueBericht_9_Akteursvielfalt_final.pdf

Kahla, F., Holstenkamp, L., Müller, J.R. & Degenhart, H. (2017), Entwicklung und Stand von Bürgerenergiegesellschaften und Energiegenossenschaften in Deutschland. Working Paper Series in Business and Law, No. 27.

Kahles, M. (2014), Ausschreibungen als neues Instrument im EEG 2014. *Stiftung Umweltenergierecht, Würzburger Berichte zum Umweltenergierecht, Vol. 6*. Retrieved from http://www.stiftung-umweltenergierecht.de/fileadmin/

pdf_aushaenge/Aktuelles/WueBericht_6_Ausschreibungen_im_EEG_2014_2014-07-16.pdf.

Kemfert, C. & Schäfer, D. (2012), Finanzierung der Energiewende in Zeiten großer Finanzmarktinstabilität. In: DIW Wochenbericht, Nr. 31, 2012, p. 3–14. Berlin: Deutsches Institut für Wirtschaftsforschung.

Klessmann, C. (2014), *Wie lassen sich Akteursvielfalt und Bürgerenergieprojekte im Ausschreibungsdesign berücksichtigen? Fachgespräch Grüne Bundestagsfraktion am 24.09.2014. Berlin: Ecofys.* Retrieved from http://oliver-krischer.eu/fileadmin/user_upload/gruene_btf_krischer/2014/3_Ecofys_Ausschreibungsworkshop_Gruene-BT.pdf.

Leuphana Universität Lüneburg & Nestle, U. (2014), *Marktrealität von Bürgerenergie und mögliche Ausweitungen von regulatorischen Eingriffen. Studie im Auftrag des Bündnis Bürgerenergie e.V. und des Bund für Umwelt und Naturschutz Deutschland e.V. (BUND).* Retrieved from http://www.enklip.de/resources/Studie_Marktrealitaet+von+Buergerenergie_Leuphana_FINAL_23042014.pdf

Mono, R. (2013), *Umsetzung der Energiewende durch Bürgerbeteiligung.* Retrieved from http://www.leuphana.de/fileadmin/user_upload/PERSONAL PAGES/_ijkl/kahla_franziska/1._Rene_Mono.pdf

Moss, T., Becker, S. & Naumann, M. (2014), Whose energy transition is it, anyway? Organisation and ownership of the Energiewende in villages, cities and regions. Local Environment. *The International Journal of Justice and Sustainability. 12/2015*, 1547–1563.

Münchmeyer, H., Kahles, M. & Pause, F. (2014), *Erfordert das europäische Beihilferecht die Einführung von Ausschreibungsverfahren im EEG? Hg. v. Stiftung Umweltenergierecht (SUER). Würzburger Berichte zum Umweltenergierecht, Vol. 5.* Retrieved from http://www.stiftung-umweltenergierecht.de/fileadmin/pdf_aushaenge/Aktuelles/WueBericht__5_Beihilferecht_Erfordernis_Ausschreibungen_final_2014-07-16.docx.pdf

Müller, J.R. & Holstenkamp, L. (2015), *Zum Stand von Energiegenossenschaften in Deutschland. Aktualisierter Überblick über Zahlen und Entwicklungen zum 31.12.2014. Leuphana Universität Lüneburg, Arbeitspapierreihe Wirtschaft and Recht Nr. 20.* Retrieved from http://www.buendnis-buergerenergie.de/fileadmin/user_upload/downloads/Studien/Studie_Zum_Stand_von_Energiegenossenschaften_in_Deutschland_Leuphana.pdf

Nestle, U. (2015), Ausschreibungen für Erneuerbare Energien: Überwindbaren Hemmnisse für Bürgerenergie? Eine wissenschaftliche Expertise von EnKliP. Im Auftrag des Bündnis Bürgerenergie e.V. Kiel. Kiel: EnKliP

Niederberger, M. & Wassermann, S. (2015), Die Zukunft der Energiegenossenschaften: Herausforderungen und mögliche Ansätze für zukünftige Geschäftsmodelle. *Energiewirtschaftliche Tagesfragen, 65(8)*, 55–57.

Ott, E. & Wieg, A. (2014), Please, in My Backyard — die Bedeutung von Energiegenossenschaften für die Energiewende. In C. Aichele & O.D. Doleski

(Eds), *Smart Market. Vom Smart Grid zum intelligenten Energiemarkt* (pp. 829–841). Wiesbaden: Springer Fachmedien

Schreurs, M. A. (2008), From the bottom up: local and subnational climate change politics. *The Journal of Environment and Development, 17* (4), 343–355.

Schweizer, P.-J., Renn, O., Köck, W., Bovet, J., Benighaus, C., Scheel, O. & Schröter, R. (2014), Public participation for infrastructure planning in the context of the German "Energiewende". *Utilities Policy. 43*, Part B, 206–209.

Sumpf, P. (2014), Energiewende und Vertrauen (Energy Transition and Trust). GAIA 23/3, pp. 287–288.

Tews, K. (2014), Europeanization of energy and climate policy: new trends and their implications for the German energy transition process. FFU-Report 03-2014. Berlin: Environmental Policy Research Centre, Freie Universität Berlin.

Tews, K. (2018), The crash of a policy pilot to legally define community energy. Evidence from the German Auction Scheme. *Sustainability 2018, 10*(10), 3397.

Trendresearch & Leuphana Universität Lüneburg. (2013), Definition und Marktanalyse von Bürgerenergie in Deutschland. Im Auftrag der Initiative „Die Wende — Energie in Bürgerhand" und der Agentur für Erneuerbare Energien. Bremen/Lüneburg: Trendresearch and Leuphana Universität Lüneburg.

Trendresearch. (2018), Eigentümerstruktur: Erneuerbare Energien. Entwicklung der Akteursvielfalt, Rolle der Energieversorger, Ausblick bis 2020. Bremen: Trendresearch.

Vogelpohl, T., Ohlhorst, D., Bechberger, M. & Hirschl, B. (2017), German renewable energy policy – independent pioneering versus creeping Europeanization? In I. Solorio & H. Jörgens (Eds), *The EU renewable energy policy: challenges and opportunities* (pp. 45–64). Cheltenham: Edward Elgar.

Verbraucherzentrale Bundesverband. (2013), *Vom Verbraucher zum Stromerzeuger. TNS Emnid Umfrage im Auftrag der vzbv.* Retrieved from http://www.vzbv.de/pressemitteilung/vom-verbraucher-zum-stromerzeuger

Walk, H. (2014), Energiegenossenschaften: neue Akteure einer nachhaltigen und demokratischen Energiewende? In A. Brunnengräber & M.R. Di Nucci (Eds), *Im Hürdenlauf zur Energiewende* (pp. 451–464). Wiesbaden: Springer Fachmedien.

Wassermann, S., Reeg, M. & Nienhaus, K. (2015), Current challenges of Germany's energy transition project and competing strategies of challengers and incumbents: The case of direct marketing of electricity from renewable energy sources. *Energy Policy 76*, 66–75.

WBGU Wissenschaftlicher Beirat der Bundesregierung Globale Umweltveränderungen. (2012), *Finanzierung der globalen Energiewende. Berlin.* Retrieved from http://www.wbgu.de/fileadmin/templates/dateien/veroeffentlichungen/politikpapiere/pp2012-pp7/wbgu_pp7_dt.pdf

Zuber, F. (2014), *Der Bürger als Treiber der Energiewende: Vom passive Konsumenten zum aktiven Gestalter der lokalen Energieversorgung? Bündnis Bürgerenergie e.V.* Retrieved from http://www.polsoz.fu-berlin.de/polwiss/forschung/systeme/ffu/forschung-alt/projekte/laufende/11_energytrans/konferenz2014/programm/2-buergerenergie-Zuber.pdf

2.6 India: dirty versus clean decentralised energy generation

Ranjit Bharvirkar

2.6.1 Introduction

Typically, in most of the developed world (and China), provision of electricity to customers has evolved from a decentralised version (e.g. a building served by a small coal-thermal generator in its basement, as seen in the early 1900s in New York city) to massive power plants connected by high voltage transmission lines serving hundreds of thousands of customers over wide swathes of areas (referred to as 'centralised' systems). It is only in the last decade – with the advent of increasingly affordable solar PV – that customer-sited electricity generation has made a comeback in the developed world.

The developing world – including India, which is the focus of this chapter – is still somewhere on this path of moving from a decentralised to a centralised electricity system. Unlike the developed world, customers in India still meet their electricity needs through a combination of the centralised power system and decentralised systems served by the private sector (e.g. diesel gensets, kerosene lanterns, lead acid batteries, gas water heaters, etc.). As per India's Central Electricity Authority, as of 31 July 2018, the centralised sector accounts for ~345 GW of installed generation capacity with an additional 60–90 GW of captive generation installed by large commercial and industrial consumers on their premises (Central Electricity Authority 2018a). Good quality and comprehensive data on the existing decentralised systems mentioned above is non-existent. It is estimated that ~250 million Indian citizens are not yet connected to the grid – accurate data is non-existent although a few organisations have started conducting surveys to ascertain this figure. The remainder of the population face highly unreliable supply as estimated by relatively crude efforts such as www.watchyourpower.org.

How to cite this book chapter:

Bharvirkar, R. 2020. India: dirty versus clean decentralised energy generation. In: Burger, C., Froggatt, A., Mitchell, C. and Weinmann, J. (eds.) *Decentralised Energy — a Global Game Changer.* Pp. 101–112. London: Ubiquity Press. DOI: https://doi.org/10.5334/bcf.g.

The adoption of these decentralised systems had nothing to do with any environmental or policy objectives. They are simply improving the customers' quality of service beyond what was being provided by the centralised system. The decentralised electricity systems are both dirty (in an environmental sense) and expensive. They are also modular and hence, customers are able to invest in them in an incremental fashion. Consequently, the quality of service as received by the customers is a function of their ability and willingness to pay. However, even the combined centralised and decentralised systems are still unable to meet all the electricity needs of millions of customers. Many are simply forced to forego some of their needs.

Historically, there have been no policies or programs to encourage decentralised systems – on the contrary, the policymakers have constantly been under pressure to improve the centralised system so that customers would not have to depend on the dirty and expensive decentralised systems.

The easy availability of clean decentralised technologies (e.g. solar PV, various battery technologies, etc.) that are now significantly cheaper than the existing dirty decentralised technologies is now altering the value proposition and allowing customers to make the substitution both from centralised to decentralised, and within decentralised from 'dirty' to 'clean'.

The performance and cost effectiveness of these new clean decentralised technologies has improved so quickly and significantly, that in many parts of the country, more than half the grid-connected load already finds it significantly cheaper – relative to the marginal retail tariff they face – to install rooftop PV in an attempt to minimise purchasing power from the grid. Bloomberg New Energy Finance (2017) has estimated that the compound annual growth rate (CAGR) since 2013 for adoption of rooftop PV systems for large consumers (typically, commercial and industrial) who face high marginal tariffs across India is ~117 per cent. The centralised grid in India may never become what it used to be in the developed world – i.e. the sole provider of reliable and cheapest electricity to all customers. In fact, India appears to be moving towards a hybrid system – i.e. a mix of centralised and decentralised systems – that is better able to meet its energy needs.

2.6.2 A brief history of systemic changes

Historically, electricity generation started with small generators serving small loads that were located next to the load itself – e.g. the Pearl Street Station in New York in the 1880s served ~400 lights and 85 customers and was literally located in the neighbourhood. As the benefits of economies-of-scale were recognised, the sizes of the generating stations grew to 100s if not 1000s of megawatts serving 100s of millions of customers. Inevitably, these stations were located away from the load centres and electricity had to be transmitted over high voltage transmission lines. And over time the small-scale neighbourhood or customer-sited generators were eliminated except for a small portion of customers that have a requirement of extremely high levels of reliability.

In most of the developed world (and China), today the main source of electricity is this centralised system (i.e. large-scale generators, high voltage transmission networks, and low voltage distribution networks) that is both reliable and affordable.[24] The use of centralised systems, driven by their low costs due to economies-of-scale has been common across the developed world outside the power sector also – e.g. natural gas, heat, water, phone/cable, etc. All of these centralised systems are extremely complex – in terms of the number of components in the supply chain that have to work well and in concert with each other – and yet they are also extremely reliable in these countries.

It is only in the last decade or so that customers have begun to adopt distributed generation – e.g. solar PV – largely encouraged through policies/programs where the objectives at least initially did not include either reliability or cost effectiveness. The objectives were numerous including but not limited to environmental protection, climate change mitigation, etc. For a detailed discussion of these objectives see Deshmukh et al. (2012). The value proposition – i.e. quality of the electric service and its cost – from the centralised system for <u>most</u> customers in the developed world (and China) appears to be no longer true across the world as cost of rooftop PV (even unsubsidised) have started edging below the retail tariffs faced by consumers – see for example, Australia and Hawaii.

India, too, embarked on a similar trajectory where the centralised system was expanded rapidly, especially, after gaining Independence in 1947. It is important to note that the data and information presented in this paragraph represents only the centralised system. The installed capacity at that time was ~1.4 GW serving a population of ~345 million (Central Electricity Authority 2018b). Today, after more than seventy years, the installed capacity has reached ~345 GW (or more than 200 times that of 1947) while the population has quadrupled.

Of order 250 million citizens in India still do not have access to electricity – in as basic a sense as having a wire reaching inside their household. Several efforts to validate the claims of 'electrification' (i.e. www.garv.gov.in) indicate that the definition of 'electrification' continues to be suspect – see for example, Patel (2016), Bansal (2016), Sharma, Josey and Sreekumar (2016), etc. Those who do have access to electricity routinely face power outages that can stretch into several hours per day (Sengupta 2016). Unfortunately, rigorous data on the exact level of 'true' electrification and quality of service is not available in India

---

[24] The exceptions in the developed world in terms of affordability are few – e.g. Australian customers are increasingly finding solar PV (including batteries) to be cheaper than the centralised system. Most of the other examples where distributed solar PV makes sense in the developed world are limited to isolated systems such as those on islands or located in remote areas.

in a comprehensive manner. Efforts such as Prayas Energy Group's Electricity Supply Monitoring Initiative (ESMI) are providing at least a glimpse about the quality of service.[25]

Rolling blackouts are – in fact – so common for most Indian citizens that in many states the local electricity distribution companies (commonly referred to as 'discoms' in India) provide formal schedules by location and time in local newspapers and online. These schedules are approved by electricity regulators and consumer advocates. Of course – as there is no way to monitor accurately, the discoms can and do deviate from these formally announced schedules. In fact, the infamous power outage in 2012 that blacked out two thirds of India simultaneously, as experienced by citizens in India, was no different in its length than the usual outages they face on a regular basis.

The two key characteristics of electricity from a centralised system in India – lack of access and when connected albeit with poor reliability – have, obviously, not resulted in citizens simply foregoing end-uses (e.g. lighting, water heating, electronics, etc.) that rely on electricity completely. After all, the Indian economy has been growing rapidly for several years – averaging 6–7 per cent annually over the last decade – that has in turn led to a rapid growth in the income albeit not uniformly across all income classes. Consequently, Indian citizens have always sought to supplement their consumption of electricity from the centralised system with a wide range of alternatives.[26] The alternatives that are most prevalent in India include:

1. Electricity: diesel generators and lead-acid batteries.
2. Lighting: kerosene lanterns.
3. Water heating: liquefied petroleum gas (LPG), wood, etc.
4. Water pumping: diesel-fired pumps.

The variations and combinations in which these alternatives are deployed by each user range widely and are driven by factors such as the requirements of the user, ability and willingness to pay, availability of alternatives, etc. These not only vary among users but also vary over time for the same user. And – most important of all – these alternatives are not even acknowledged by the centralised system (i.e. the local electricity distribution company and policymakers) let alone being encouraged and supported. On the contrary – whenever there are periods of the centralised system being able to provide electricity at high level of reliability, the investments of users in these alternatives get completely stranded without any possibility of recourse.

---

[25] http://www.watchyourpower.org/the_initiative.php.

[26] And this observation applies to other sectors also – e.g. natural gas provided in cylinders in the form of liquefied petroleum gas, water supply service quality is similar to electricity (i.e. unreliable and poor), and phones/cables have been usurped by distributed technologies such as satellite dishes and cell phones.

2.6.3 *Potential for decentralised energy*

A thriving marketplace exists all over India that provides these alternatives to individual customers under a wide range of contractual arrangements such as leasing diesel gensets, outright purchase, and others. However, comprehensive data of a high quality about this marketplace is not available for India. Not only is basic data – e.g. sales of diesel generators –not readily available, there is not much data available about the usage of these alternatives (e.g. hours of use for a typical lead acid battery). Market research firms provide estimates – see for example 6WResearch (2016). To be sure, lack of data does not imply that these alternatives do not exist. In reality, the Indian end-user demand for electricity has always been met through a combination of centralised system, and decentralised alternatives in sharp contrast to the situation in the developed world (including China). The extent of the decentralised system is not comprehensively quantified although numerous case studies have been done that provide a useful qualitative picture.

From a cost-effectiveness perspective relative to the centralised system – most of the alternatives listed above are massively more expensive than the centralised system. For example, running a diesel generator is twice as expensive as even the most expensive retail tariff for electricity from the centralised system. Similarly, kerosene lanterns are among the most expensive ways of providing lighting. However, given the unreliability and often availability of the centralised system, the user is – in fact – assessing the cost effectiveness of the alternatives relative to having to forego the service (e.g. lighting) in its entirety.

It is in this context that one has to examine the role of new technologies such as distributed solar, batteries, more efficient equipment (e.g. LEDs), and others and assess the factors that influence their adoption. In India, distributed solar PV systems are – fundamentally – not competing with electricity from the unreliable Indian centralised system but with the substantially expensive distributed alternatives (e.g. diesel generators, etc.) that have been historically used to supplement the centralised system. Initially, solar PV was significantly more expensive than these alternatives.

From a meeting of the reliability needs of a consumer – the comparison between the two sources of distributed generation (i.e. rooftop PV and diesel) is not straightforward for two reasons:

- solar PV is available only during day-time and if there is sufficient space available for its installation while diesel generators are available on demand and have a relatively small footprint;
- the upfront (or fixed) costs of the current alternatives (e.g. diesel generators or kerosene lanterns) are relatively low with high variable costs in sharp contrast to that of solar PV – as availability of cheap credit is limited for most consumers.

Yet – the value proposition has started becoming so compelling that even the key manufacturers of diesel generators in India (e.g. Jakson, Sukam, Kirloskar, etc.) have added solar PV to their portfolios – whether independently or hybridised with diesel generators (Paul 2015). For example, Jakson Inc. – one of the main diesel generators in India – forecast in 2014 that within three years half of their sales would be from solar PV (Pearson 2014).

Not only is solar PV successfully competing against alternatives such as diesel generators, it has also now started competing against the retail tariffs faced by commercial and industrial (C&I) customers in India for the electricity they purchase from the centralised system – see Prayas Energy Group (2017) and Bloomberg New Energy Finance, 2017. The high retail tariffs for C&I customers in India is a consequence of the distinct policy environment which is discussed in more detail in the next section. For now, we simply note that these high retail tariffs are unlikely to decrease in the future – see Central Electricity Authority (2018) for trend in Average Revenue Recovered (i.e. average retail tariff) by discoms.

The Indian national and state policymakers have in recent years announced and initiated the implementation of several policies to support distributed solar PV. Forty GW out of a total of 100 GW of solar PV that is set as the national target for 2022 is allocated to rooftop PV. However, discoms have started confronting the distinct possibility that the early adopters are likely to be their larger and wealthier consumers. These consumers provide the cross-subsidy that sustains the utility cash flow. If the utility sales to these large consumers falter, then there would be an immediate and significantly adverse impact on the utility financial situation. Consequently, utilities in many parts of India have started resisting the growth of rooftop PV that has led to rumours that the 40 GW policy target may end up being revised downward significantly. However, there is not much anyone can do when consumers install behind-the-meter rooftop PV systems that simply and passively offsets their consumption from the grid. From the discom's perspective, this appears similar to load that has vanished somewhat akin to but a potentially significantly larger impact than that of energy efficiency or demand response.

2.6.4 The drivers of decentralised energy

These circumstances – i.e. unreliability, increasing costs of the centralised system, and cost effectiveness of rooftop PV relative to retail tariff for a growing number of its customers – are unlikely to change in the future creating an attractive market for distributed solar PV without any policies and programs designed to promote it. In this section, we discuss in more detail why these circumstances are likely to persist in the future.

Historically, electricity has been one of a number of public goods for which the government was seen to be the sole provider. On paper – there are specific departments for each type of infrastructure – e.g. the state electricity distribution companies, municipal transport and housing authorities, centrally owned railways, etc. However, from the citizens' perspective, it was the state government that was supposed to provide them with electricity, just as it was meant to provide roads, healthcare, education, water, sanitation, public safety, etc. The electricity ratepayers of the government-owned utility also are the voters that state governments care about at election time. Issues like the cost of electricity to consumers not only come up in regulatory proceedings but also routinely show up in politics – see for example Dubash, Kale and Bharvirkar (2018).

Within India's federal division of power, state governments have the lion's share of authority in the electricity sector. After independence in 1947, the hundreds of small private utilities that generated and distributed electricity throughout the subcontinent were gradually subsumed into State Electricity Boards that were wholly owned by the respective state governments. In the first few decades post-Independence, the majority of the investment in the power sector was made by the state governments.

Over the last 15–20 years, these SEBs were unbundled into generation, transmission, and distribution companies. The national government has been attempting to introduce amendments to the 2003 Electricity Act with the goal of introducing full retail competition. However, there has been sustained resistance from the state governments to this step.

Significant investment from the central government and private sector has been made in the generation-side of the power sector, even today, state governments own most of the electricity distribution companies in India.

The world of the state government-owned distribution companies consists of trying to both expand massively its electricity distribution network (remember the 200–300 million customers with no access and the fact that India's population is still growing rapidly) and maintain the existing one, which has high wear-and-tear and as customers need increasing amounts of electricity to power air conditioners and other conveniences. All of this must take place in a context in which the retail tariff at which electricity can be sold to customers (also voters!) is fundamentally subject to political constraints and therefore remains low.

Given that India is still a poor country with many competing demands for limited government resources, it is difficult for state governments to come up with direct subsidies to keep the price of electricity low for its citizens, particularly the electorally significant agricultural users and low-income households. The alternative, which most state governments have chosen, is to achieve that outcome by charging richer and larger customers, who are typically commercial and industrial users, higher-than-cost tariffs while charging low income and agricultural customers tariffs that often approach zero.[27] This cross-subsidy

[27] From a techno-economic perspective, the cost of providing service to large (e.g. C&I and urban residential) customers is the least due to economies of scale and density

system—in effect, based on the same principle as a progressive income tax policy—runs into obvious limitations. Beyond a certain point, richer and larger customers will either invest in their own electricity supply and/or leave the service territory completely. Either circumstance worsens the already difficult situation of the state-owned distribution utilities. Consequently, the distribution companies are constantly trying to strike a balance between their mandate to provide reliable power to all customers (also voters!) while keeping the costs at a level that are deemed politically acceptable.

From a broader perspective – i.e. how Indian citizens have received a wide range of services similar to electricity – the experience has been that the centralised system has consistently been unable to keep up with the demand. Consequently, Indian citizens have been forced to develop alternatives to the centralised systems in many aspects of their lives. Barring some solely centralised and publicly provided services (e.g. national security, roads, railways, etc.), decentralised and privately provided services supplement (even substitute) the centralised system in case of services such as education, transportation, electricity, water, healthcare, food, and others.

The service most analogous to electricity is water. And this example, too, is telling. Unlike in western countries, most Indian cities have not been able to supply their citizens with 24-7 water through a centralised pipeline system. Consequently, almost all households and businesses have water storage tanks. Some households have multiple. Many households have private 'bore wells' that run deep underground to access subterranean aquifers; others hire companies to deliver water in private tankers. Similarly, since the quality of the water is poor, households rely on a host of purification technologies installed and operated at the point of consumption.

As the costs of decentralised systems in the electricity domain continue to decrease – i.e. rooftop PV (analogous to bore wells in the water infrastructure above) and batteries (analogous to water storage tanks in the water infrastructure above) – it is quite possible that the electricity system would also evolve into a hybrid system similar to the water supply system. The problem is that the current water system (and the future potential electricity system) is not optimally designed, implemented, and operated in this hybrid form – but has simply evolved in a haphazard manner. A more thoughtful approach that actively incorporates both forms of centralised and decentralised systems in order to minimise the overall costs to society would be beneficial. For example, the distribution utilities may want to formally incorporate the decentralised systems (both spatially and temporally) while designing and operating their distribution grid infrastructure and making their wholesale level procurement decisions. Retail tariff designs could be considered that provide appropriate signals

---

and highest for the small (e.g. rural and agriculture) customers. Consequently, the cross-subsidy mechanism is, especially, unattractive for large customers.

to consumers as they determine which decentralised systems to invest in. Currently – the distribution utilities have either ignored the existence of the ad hoc decentralised systems or resisted them outright.

Acknowledging that there are important variations across states in India, we can nevertheless make some general observations about the prevailing equilibrium: distribution companies are starved of capital to expand, an Indian population equivalent to that of the United States remains without electricity connections, and everyone else has a poor quality of service. However much the bureaucrats, engineers, and politicians who work in India's central and state governments might yearn to achieve their ultimate objectives of 24-7 power for all through the centralised system, they are constrained by political and economic conditions. Consequently, the status quo for the quality of the service provided by the centralised system is unlikely to change in the near term.

2.6.5 Outcomes

As discussed in the previous two sections, the existence of decentralised alternatives in India, albeit non-RE-based, long pre-dates the current interest in the United States and European Union for supporting decentralised RE-based alternatives. The circumstances that sustain the existence of these decentralised alternatives are unlikely to change substantively in the near term. And as the cost of RE-based decentralised technologies drops below that of existing decentralised technologies and the cost of electricity from the centralised grid, the uptake of RE is likely to continue. This uptake of decentralised RE is likely to take place independent of government policies and programs. For example, Prayas Energy Group (2017) has estimated that when the levelised cost of rooftop PV system reaches Rs. 5 per kWh in India – in nine major states in India, more than 50 per cent of the non-agricultural sales would find it cost effective to start switching to rooftop PV. In five of those nine states, the proportions is more than 70 per cent. As per BNEF (2017), all across India, prices for rooftop PV discovered through auctions have already sunk well below Rs. 4 per kWh with the minimum prices being observed as low as Rs. 2.38 per kWh.

Appropriately designed and implemented government policies and programs could expedite this uptake significantly. While poorly designed and implemented policies and programs may at best not influence the rate of uptake but at worst, is likely to slow it down. In this section, we discuss the implications of the existing policies/programs and suggest some approaches worth considering.

Most of the policy mechanisms currently being implemented or considered in India are adapted from those in the developed countries (e.g. the United States) where the value proposition for decentralised technologies is fundamentally different than that in India. To reiterate – decentralised technologies in India compete with an unreliable (sometimes non-existent) and an increasingly expensive centralised system. Unlike a typical customer in the developed world that expects the centralised system to be extremely reliable and affordable, the expectation of a typical Indian customer is to be able to pay for the level of reliability they seek and can afford. Indian policies can better take into account these two unique conditions in order to develop approaches that are better suited for the Indian context.

One possibility is a national-level program similar to the successful LED-distribution program (http://www.ujala.gov.in) for solar PV panels, batteries, and inverters could be implemented that would streamline the availability of those products significantly while ensuring their quality at the lowest cost possible (through aggregation of demand and negotiating with manufacturers). Further, storage in the form of electricity (i.e. batteries and inverters) is not necessarily the only way to store useful energy. Thermal storage systems such as ice for space cooling and water heaters are a mature technology that could be deployed where appropriate. Other distributed systems such as geothermal heat-pumps can also be considered as the technology matures.

There would be no subsidies involved as the program is designed in the form of a loan that is recovered through an on-bill financing mechanism in partnership with the local utility. A wide range of sizes of technology (i.e. make full use of the modularity of the technology!) could be available to customers through this program for procurement either individually or in the form of cooperative arrangements. In conjunction with this technology dissemination program, retail tariffs could be re-designed to ensure that the costs/benefits are shared fairly between the utility and the customers. Both the approaches described here don't' rely on large financial commitments from the government and instead attempt to make full use of the existing demand for these technologies. For programs aimed at specific end-uses – e.g. agricultural pumps – the design must ensure that negative indirect impacts (e.g. water/land overuse) are minimised if not avoided through the use of a comprehensive portfolio of policies that go beyond the power sector (e.g. crop selection, water-use efficiency, etc.).

In India, the centralised power system will continue to provide a certain level of reliable electricity supply at prices established through a process that remains essentially political.

Outside of this bound, though, customers will continue to enhance their supply from the outside, where options have evolved from no electricity to lead-acid batteries and diesel generators, and now to solar PV and advanced batteries. In the absence of a more thoughtful approach to designing and operating the grid (e.g. retail tariff design) – the hybrid form of the power sector will evolve in a chaotic manner thereby imposing needless costs on the society. The goal of the policymakers – then – is to find ways to ensure their thinking about the future power sector is not largely restricted to the centralised system with decentralised systems a mere after-thought at best or outright nuisance at worst – but one of the important components of the power sector that can yield even lower costs for all.

2.6.6 References

6WResearch. (2016), *India diesel genset market (2018-2023), GWresearch, June 2017.* Retrieved from http://www.6wresearch.com/market-reports/india-diesel-genset-generator-market-2018-2023-forecast-by-kva-rating-verticals-commercial-and-industrial-regions-competitive-landscape.html

Bansal, S. (2016, March 26), On paper, electrified villages — in reality, darkness. *The Hindu*. Retrieved from http://www.thehindu.com/opinion/op-ed/On-paper-electrified-villages- percentE2 percent80 percent94-in-reality-darkness/article14176223.ece

Bloomberg New Energy Finance. (2017), *Accelerating India's clean energy transition: the future of rooftop PV and other distributed energy markets in India.* Retrieved September 16, 2018, from https://data.bloomberglp.com/bnef/sites/14/2017/11/BNEF_Accelerating-Indias-Clean-Energy-Transition_Nov-2017.pdf

Central Electricity Authority. (2018a), *Executive summary on power sector.* Retrieved September 16, 2018, from http://cea.nic.in/reports/monthly/executivesummary/2018/exe_summary-07.pdf

Central Electricity Authority. (2018b), Growth of electricity sector in India from 1947-2017. Retrieved September 16, 2018, from www.cea.nic.in/reports/others/planning/pdm/growth_2017.pdf

Deshmukh, R., Bharvirkar, R., Gambhir, A. & Phadke, A. (2012), Changing sunshine: analyzing the dynamics of solar electricity policies in the global context. *Renewable and Sustainable Energy Reviews, 16* (7).

Dubash, N., Kale, S. & Bharvirkar, R. (2018), Mapping power: the political economy of electricity in India's states. Oxford: Oxford University Press.

Patel, D. (2016, September 14), All villages electrified, but darkness pervades. *Indian Express*. Retrieved from http://indianexpress.com/article/india/india-news-india/electricity-in-india-villages-problems-still-no-light-poverty-3030107/

Paul, C. (2015, January 7), Policy changes and private enterprise to boost greener India with cleaner fuel. *Forbes India*. Retrieved from http://www.forbesindia.com/printcontent/39311

Pearson, N. (2014, May 12), Jakson says solar to match diesel-generator sales by 2017. *Bloomberg*. Retrieved from https://www.bloomberg.com/news/articles/2014-05-12/jakson-expects-solar-to-match-diesel-sales-by-2017

Prayas Energy Group. (2017), *The price of plenty: insights from "surplus" power in Indian states.* Retrieved September 16, 2018, from http://www.prayaspune.org/peg/publications/item/335-the-price-of-plenty-insights-from-surplus-power-in-indian-states.html

Sharma, A., Josey, A. & Sreekumar, N. (2016), Lessons for rural electrification from a weaving village. *India Together*. Retrieved from http://indiatogether.org/lessons-for-rural-electrification-from-a-weaving-village-government

Sengupta, D. (2016, July 26), Contrary to government's claims, small towns, rural areas still suffer from power outages. *Energyworld*. Retrieved from https://energy.economictimes.indiatimes.com/news/power/contrary-to-governments-claims-small-towns-rural-areas-still-suffer-from-power-outages/53393538

2.7 Italy: network costs versus decentralised system

Michele Gaspari and Arturo Lorenzoni

2.7.1 *Introduction*

The diminishing economies of scale and the gradual cost reductions of renewable technologies make distributed generation (DG) the most convenient alternative for new electricity generating installations (Lorenzoni 2014; Lorenzoni 2015; REN21 2016). The establishment of a decentralised electricity market is essential for compliance with the decarbonisation commitments of the Paris Agreement (UNFCCC 2015; IEA 2016), even if in the framework of European targets (European Commission 2014) each member state may implement different instruments for the transition to sustainable energy (Kuzemko et al. 2016). The Italian market represents a significant case for the analysis of the evolution towards low-carbon electricity systems as a remarkable increase in renewable capacity installed in the last six years was possible within a market characterised by a real liberalisation process even with a large share of the former incumbent Enel (ARERA 2018). However, whilst Italy has already met its 2020 decarbonisation goal for the electricity sector, some policies currently in place may be detrimental for the further development of decentralised energy resources (DER): in order to revert this trend, the draft National Energy and Climate Plan presented in December 2018 set ambitious goals to further increase decarbonisation (Ministero dello Sviluppo Economico, Ministero dell'Ambiente, Ministero delle Infrastrutture 2018).

This chapter provides an overview of the most significant regulatory and market aspects that have enabled Italy to meet its target, and of the barriers that are hampering further actions. The second section describes the institutions and characteristics of the Italian electricity market as well as the most important decarbonisation policies; the third section analyses the outcomes of

---

How to cite this book chapter:
Gaspari, M. and Lorenzoni, A. 2020. Italy: network costs versus decentralised system. In: Burger, C., Froggatt, A., Mitchell, C. and Weinmann, J. (eds.) *Decentralised Energy — a Global Game Changer.* Pp. 113–132. London: Ubiquity Press. DOI: https://doi.org/10.5334/bcf.h.

such policies. The fourth section focuses on the main governance drivers and barriers for decentralisation, and the fifth section provides concluding remarks.

2.7.2 The impact of European policies on the decentralisation of the Italian electricity market

From nationalisation to liberalisation: a short description of institutions and market players

In 1962 the per-capita electricity consumption in Italy was still lower with respect to other comparable countries; the centre-left government decided to nationalise the industry and to establish the single vertically integrated utility Enel, in order to give access to every consumer under the same conditions, and to sustain economic growth (Ranci 2014).

The establishment of the National Regulatory Authority (NRA), thirty-three years later, to promote competition, efficiency and transparency whilst maintaining a high quality of supply, was a precondition for the liberalisation of the market (Legge 481/1995). According to the European Directive 1996/92/CE, the first step to promote the participation to the free retail market was to give access to 'eligible consumers'; the Italian decree (D.Lgs. 79/1999) declared 'eligible' those consumers with a decreasing annual consumption threshold (from 100 GWh down to 0.1 GWh from 1999 to 2007), while the others were supplied in the regulated captive market, where supply conditions were defined by the NRA. This hybrid solution was unstable: since July 2007 everyone has access to the free retail market; however, the captive market, where a single buyer ensures electricity supply, is still available and serves the majority of households (ARERA 2018). The wholesale market is managed by Gestore dei Mercati Energetici (GME), which sets criteria for neutrality, transparency, and competition amongst producers: Terna is the transmission system operator.

Enel remains dominant across supply and distribution: with respect to distribution networks, in 2017 Enel Distribuzione (currently e-Distribuzione) managed 85 per cent of total electricity supply volumes (ARERA 2018). Enel is also the largest supplier in the retail market (85.4 TWh): a deliberation from the NRA is stricter in terms of the obligation of functional unbundling (ARERA 2015d). In general, the Italian territory is provided with scarce natural resources and fossil fuels are not available: the dependence from foreign sources is 75 per cent, with oil and gas still accounting for around 60 per cent of the energy mix (Unione Petrolifera 2016).

With respect to the electricity market agenda the Italian Government has acted as a policy-taker, building its policies according to EU priorities (Font 2002; Clò 2014). The same approach has been adopted with respect to decarbonisation commitments to be implemented in Italy after the EU signature of the Paris Agreement (UNFCCC 2015).

Table 3: Electricity generation (GWh) by renewable energy sources in Italy.

	2013	2014	2015	2016	2017	2018**
Hydro	52,773	58,545	45,537	42,432	36,199	49,28
Wind	14,897	15,178	14,844	17,689	17,742	17,492
Solar	21,589	22,306	22,942	22,104	24,378	22,653
Geothermal	5,659	5,916	6,185	6,289	6,201	6,08
Bioenergy*	17,09	18,732	19,396	19,509	19,378	19,219
TOTAL	**112,008**	**120,677**	**108,904**	**108,023**	**103,898**	**114,724**
Gross National Consumption	330,043	321,834	327,94	324,969	331,765	332,849
RES/Gross National Consumption	**33.94%**	**37.50%**	**33.21%**	**33.24%**	**31.32%**	**34.47%**

*Bioenergy includes: solid biomass (including the organic fraction of municipal solid waste), biogas, bioliquids, and biomethane.

**Provisional estimations.

Source: GSE, 2019a.

The governance of decarbonisation policies

The generous incentive schemes for renewable energy sources (RES), the cost reductions of low-carbon technologies and the flattening electricity demand due to the economic crisis have made possible that sustainable sources represented around one third of the Italian electricity generation mix in the last 6 years, as can be seen in Table 3 (GSE 2019a).

Since 1991 the Italian law declared RES projects of 'public interest' and 'public utility', and the related works 'urgent' and 'not deferrable' (Legge 9/1991). After the Directive 2001/77/EC, Italy promoted a policy (D.Lgs. 387/2003) that simplified the permitting process for these facilities. Gestore dei Servizi Energetici (GSE) is the legal entity in charge of managing the incentives for RES and purchases electricity from these generators.

The main support schemes adopted by GSE are:

- **Tariffa Onnicomprensiva (feed in tariff):** for renewable generators (excluding PV and including wind, hydro, bioenergy) entered into operation before 31 December 2012 with a capacity installed up to 1 MW (200 kW for wind); it envisions a fixed amount for each kWh produced, differentiated according to the source, for 15 years.
- **Green Certificates (GC):** for net electricity produced by RES facilities entered into operation before 31 December 2012; from 1 January 2016 GC are replaced by feed-in-premium incentives until the end of the right to obtain GC (20 years).

- **Conto Energia, incentive scheme for PV, which comprises:**
  - A feed-in-premium scheme for PV projects entered into operation before 26 August 2012.
  - A scheme for PV plant entered into operation from 27 August 2012 to 6 July 2013: feed-in tariff for projects with a capacity installed lower than 1 MW_p and feed-in-premium for larger units, with a prize for net self-consumed electricity (GSE 2016).

The diversity of instruments and dates highlights the fact that the Italian regulatory framework has not provided reliable instruments for the promotion of DER. The policies to support PV provide a particularly interesting story in these terms, with frequent and sometimes random changes of rules. For example, support for PV started in 2005, and the incentive was then modified in 2007 (D.M. 19 February 2007), in 2010 (D.M. 6 August 2010; Legge 129/2010), in 2011 (D.M. 5 May 2011), and finally in 2012 (D.M. 5 July 2012). During these years, the cost of installations decreased faster than the premiums and in 2011 nearly 10 GW were installed. In 2012 the Government passed a decree to end support for PV as soon as the overall cost of the programme reached €6.7 billion/year.

Other very important support schemes in relation to the development of a decentralised system are 'Scambio Sul Posto' and 'Ritiro Dedicato' (Nextville 2013). 'Scambio sul Posto' or net metering (ARERA 2012), is a commercial agreement with GSE valid for low-carbon units up to 200 kW: the electricity generated by an on-site installation and injected into the grid can be used to offset the electricity withdrawn from the grid itself. A total of 524,600 users have adopted this method of net metering, a total capacity of 4.5 GW (ARERA 2016a). Since 2009 this has been based on market values: users pay the total amount for their consumption and in return receive a fair contribution set at retail market prices for the electricity produced. Until 2012 this scheme was compatible with other incentives, but this is no longer the case. 'Ritiro Dedicato' (Simplified Purchase and Resale Agreement) (ARERA 2007) is a simplified formula for low-carbon facilities under 1 MW of capacity. Producers sell the electricity generated to GSE instead of selling it through bilateral contracts or directly on the wholesale market; they are remunerated with guaranteed minimum prices, while larger units receive the average monthly price set on their zonal wholesale market. 51,119 plants adopted this scheme voluntarily, a total capacity of 11.6 GW (ARERA 2016b).

In spite of the numerous regulatory turnarounds, Italy successfully complied with its decarbonisation targets, reaching 69 per cent of its target in 2011: at the end of 2015 the reduction in CO_2 emissions was 34 per cent higher than the 2020 target (ENEA 2016a). With the steady growth of the last decade, RES gained a central role in the energy sector, in both operation and policy. This was at the expense of fossil thermal electricity generation, which accounted for 82.6 per cent in 2007 and 56 per cent in 2014; electricity-related

CO_2 emissions amounted to 591.1 gCO_2 per kWh in 1990 falling to 323.6 gCO_2 per kWh in 2014 (ENEA 2016b).

National/regional policies and geographical dimension

In the Italian Constitution of 1948, energy production, transmission and distribution became state competencies, but with a constitutional reform in 2001 (L. Cost 3/2001) a division of legislative powers made the subject a matter of concurrent competencies between national and regional entities: the Italian Government establishes the general principles of the sector, complying with the supranational EU framework, while regions legislate within their territories according to such principles. The aim of this configuration was to relieve the state from an excess of tasks and to simplify the administrative procedures in the sector. A constitutional reform, which among other objectives, aimed to give back exclusive responsibility to the state in the energy field (D.D.L. Cost 2016), was rejected following a referendum in December 2016.

The geographical dimension is significant not only in terms of legislative competence: in general, northern regions consume more electricity and are characterised by a larger number of installations from RES (including hydro). Southern regions are affected by lower consumption and higher prices for electricity, mainly due to bottlenecks in the transmission grid: it often happens that RES generation is higher than total load (ARERA 2016a). Figure 14 illustrates the geographical distribution of DG in terms of number of substations involved for more than 5 per cent of total time in power flow inversions.

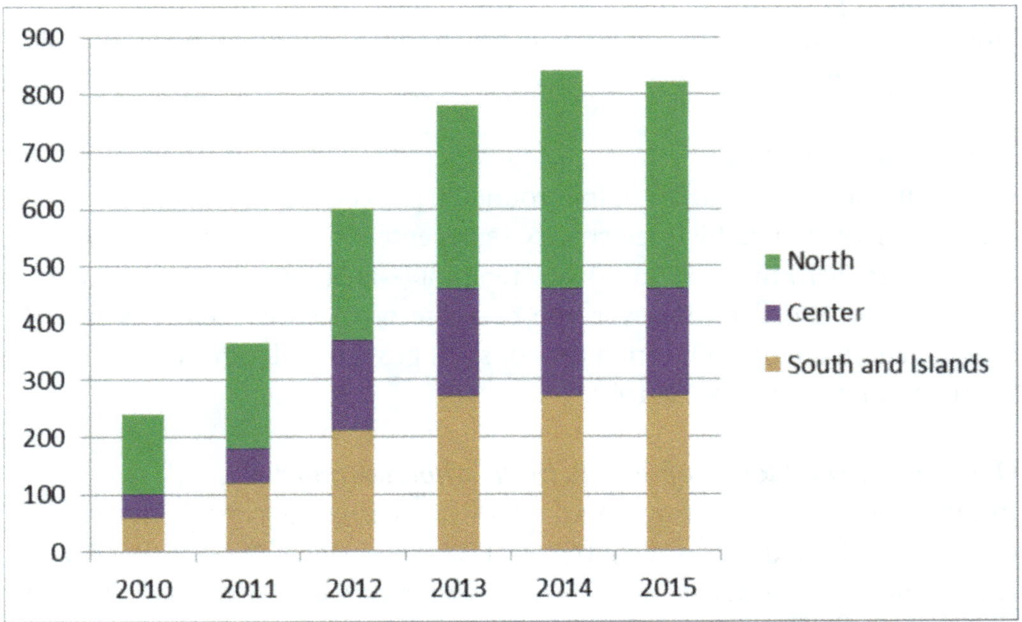

Figure 14: Number of HV/MV substations with power flow inversions > 5 per cent of total time.

Source: ARERA (2016a).

Importance of consumers in policy debate

In the last few years, the NRA devoted increasing attention to unfair commercial activities from suppliers, mainly with regards to billing processes. In general, consumer unions complain about high bills and lack of transparency from suppliers: from 2011 to 2013 the number of complaints increased from 335,000 to 500,000, 70 per cent of which related to households' contracts (Federconsumatori 2014). The liberalisation also raised concerns related to energy poverty for low-income users and a social bonus was made available for households in need (D.Lgs. 102/2014, ARERA 2015c). In spite of this, energy policy is still dominated by the supply-side operators and the diffusion of new contractual relationships based on DG hardly finds proper support from NRA.

2.7.3 The status of decentralisation in the Italian electricity market

In 2017, the Italian gross domestic product slightly increased and the electricity demand followed the same dynamics (provisional results account for +2 per cent with respect to the previous year). Despite this signal, and as a result of improved efficiencies in the system and reduction of demand from energy-intensive sectors, the electricity demand stabilised at the same level as in 2007 (319 TWh) (Ministero dello Sviluppo Economico 2018). The electrification of heat and transport, proposed in some decarbonisation scenarios, has not yet impacted on demand, as electricity consumption still accounts for approximately 20 per cent of total final energy consumption. This shrinking demand pattern does not by itself facilitate the full transition to a sustainable system, with the sector still relying on centralised regulation and managing overcapacity.

The contribution of RES

PV and hydro together account for around 70 per cent of total installed renewable capacity (in 2008 hydroelectricity represented 95 per cent of total RES in Italy). From 2010 to 2015, 23 GW of renewable facilities were installed, with nearly 20 GWp of PV; however, the decrease of incentive schemes in 2013 brought about an interruption in the constant growth of decentralised sources, as seen in Figure 15 (GSE 2019b).

Overcapacity and low profitability in the wholesale market

In 2004 the Italian electricity market (IPEX) started as a pool (central dispatch), and allowed bilateral contracts. IPEX is managed by GME and it actually entails a spot electricity market (MPE), a forward electricity market (MTE) and a platform for physical delivery of financial contracts. MPE is currently divided into three specific segments: day-ahead market (MGP); intra-day market (MI); ancillary services market (MSD, operated by Terna). MGP is a zonal market, with the particularity of a single price on the consumer side (PUN, the

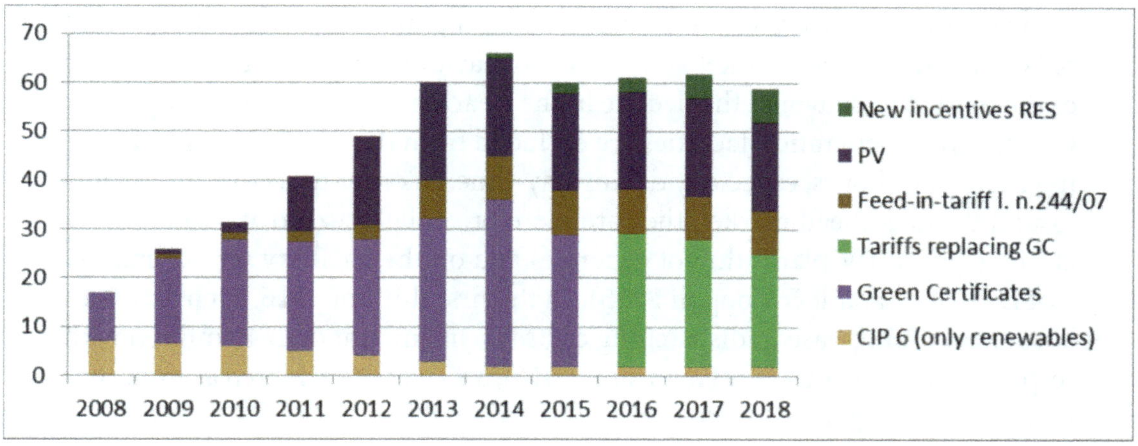

Figure 15: Subsidised electricity (TWh) generated from RES, according to incentive instrument.

Source: ARERA (2018).

weighted average zonal price) and a zonal price whenever congestion rises on the supply side. In general, the average price in the IPEX is higher with respect to other European countries (except France) because of a market largely based on gas supplies.

The large share of RES impacted the operation of IPEX, with a negative spark spread for combined cycles occurring in many months since 2012 and negative results for most of thermal generators.

Table 4: Comparison of prices available on wholesale markets in Europe.

YEARS	IPEX (Italy)	EPEX (Germany)	North Pool (Northern Countries)	OMEL (Spain)	EPEX (France)
2004	51.60	28.52	28.91	27.93	28.13
2005	58.59	45.97	29.33	53.67	46.67
2006	74.75	50.78	48.59	50.53	49.29
2007	70.99	37.99	27.93	39.35	40.88
2008	86.99	65.76	44.73	64.44	69.15
2009	63.72	38.85	35.02	36.96	43.01
2010	64.12	44.49	53.06	37.01	47.50
2011	72.23	51.12	47.05	49.93	48.89
2012	75.48	42.60	31.20	47.23	46.94
2013	62.99	37.78	38.35	44.26	43.24

Source: RSE (2015).

In general, GSE operates on IPEX as a non-programmable RES collector and bids in the day-ahead market at zero; this behaviour drives marginal units out of the market and favours the decline in the clearing wholesale electricity price. Conventional generation facilities are excluded from the merit order in a growing number of hours, especially during day-time. In order to recover the profits lost on the day-ahead market, they are therefore obliged to bid at higher prices at night (when PV plants do not generate) and on the ancillary service market (MSD) (Clò, Cataldi & Zoppoli 2015). In the first third of 2016, such activities brought to an increase in dispatching costs for an amount of €745 million with respect to the previous year (Biancardi 2016), and were considered as improper manipulations by the regulator.

Considering these aspects, the establishment of a proper capacity market would guarantee an adequate generating capacity to meet expected consumption and reserve margins (ARERA 2015f). The approved scheme (D.M. 30 June 2014) will replace the transitory system in force since 2004, which was structured as capacity payment, and entails a mechanism according to which producers will receive remuneration for the generated capacity that they make available. The final approval of the capacity market design was planned for the end of 2017, but was delayed.

Towards full liberalisation in the retail market

Twenty years have passed from the beginning of the liberalisation process, but still the majority of households (58 per cent) purchase electricity according to the regulated price. However, this trend is changing and the number of households accessing the competitive retail market is increasing: in 2017 most supply contracts for new consumption units were signed according to the liberalised framework (ARERA 2018). On average, families in the captive market consume less than families supplied in the free market (1.852 kWh/year against 2.119 kWh/year), because larger consumers are more likely to search for cheaper options; nonetheless, prices on the free competitive retail market can be higher than regulated ones, often because these offers include forms of electricity-related services.

Reforms (ARERA 2015c) are in place to encourage this shift to the free market model and reduce the role of the single buyer: the end of the captive market is expected by June 2020. However, in the presence of significant informative asymmetries, the single buyer is still a useful benchmark for the market.

In general terms, the energy bill structure is composed of 4 sections (ARERA 2018): in 2017 the energy-commodity cost section, which is related to the wholesale market price of the energy and the commercial margins of the retailer/reseller accounted for 44%, while taxes and network costs represented 13% and 20% of the whole amount, respectively. The second most significant portion of the bill is represented by general system charges, which cover the costs that the system bears for the incentive schemes to renewables and high-efficiency cogeneration, as well as other costs which are generally referred to

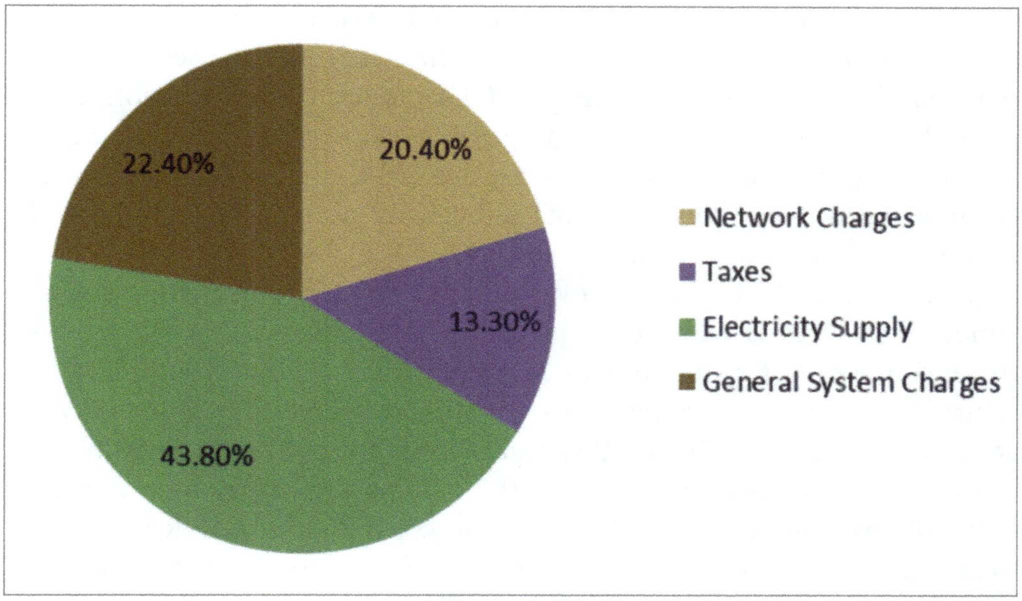

Figure 16: Costs structure for a household consuming 2,700 kWh per year and with a withdrawal capacity of 3 kW in the captive market.
Source: ARERA (2018).

the electricity system (including the nuclear plants decommissioning, the support to national railways electric systems, the bonus for fuel poverty, etc.). This latter portion accounts for 22 per cent of the total bill, and nearly 80 per cent of such overheads are devoted to incentive schemes for RES. Each of these cost items can be split into a 'trinomial structure' already presented above: a fixed value (€/connection point/year), a power capacity value (€/kW/year, based on the power capacity of the connection point), and a volumetric variable value (€ per kWh), as seen in Figure 16.

2.7.4 The Italian regulation for distributed energy resources

The previous paragraphs have already described the Italian electricity market, which complies to the requirements of the European Directives (unbundling and Third Party Access): with respect to the grid connection of renewable generators, network operators are obliged to connect them at a cost that is proportional to the distance from the connection point. However, the owner of a renewable energy plant does not have alternative solutions to self-consumption or sale to the grid. The direct sale of electricity to other consumers, as well as load aggregation, is forbidden, with the exception of the one-to-one supply under SEU (Sistemi Efficienti di Utenza, efficient user system) scheme, a regulated business model for electricity sales from DG, described below.

The biggest concern for the Italian regulator is related to the payment of system costs when a growing number of consumers are becoming self-producers, reducing the withdrawal from the grid (and the related participation to grid costs), but taking full advantage of grid services. As a matter of fact, in the current framework, innovated by the Decree 244/2016, all the electricity consumers are obliged to pay network tariffs and general system charges only on the energy withdrawn from the public network.

The current system for the recovery of the costs for transmission and distribution services is based on a price cap mechanism on operating costs (to encourage cost reductions in managing infrastructure) and a rate of return mechanism on capital costs (to stimulate investments for network adequacy) (Legge 290/2003). This regulation until 2012 made possible significant investments in transmission (€7 billion) and distribution networks (€18 billion) (Polo et al. 2014). In a recent consultation the NRA proposed to introduce an approach based on total costs (*totex*) for the remuneration of services, suggesting also the aggregation of smaller DSOs (ARERA 2015a). In general, the NRA itself recognises the need to establish mechanisms to coordinate the strategies of generation facilities and to take advantage of the flexible demand.

Governance barriers for distributed energy resources

Households' electricity tariff reform: displacing efficient consumers and on-site generators

As highlighted in the previous section, only around 40 per cent of the total costs included in the electricity bill is exposed to market competition. This structure (Ranci 2014) was aimed to keep the electricity costs for small capacity withdrawals for households (3 kW) as low as possible, without being affected by volatility. Tariffs were given a progressive structure for the recovery of network costs (transmission, distribution and measurement) and overheads: the charges grew proportionately to consumption and therefore larger consumers were burdened with the recovery of fixed costs.

Under the current scheme (ARERA 2015b), a growing portion of the bill is due to a fixed charge, and the final price is more cost-reflective. According to the NRA, the new tariff structure is favourable for consumers that put in place energy savings initiatives, because the energy component still represents at least 70 per cent of the total bill; on the other side, this scheme also supports the development of the electric options for transport and heat (electric vehicles and heat pumps), see Figure 17.

However, even if larger consumers are granted with economic savings (€164/year for families consuming 4,000 kWh/year), this reform seems detrimental for families with an annual consumption which is lower than the average and that own an on-site generation plant. Moreover, the adoption of information and control technologies in energy management to reduce electricity loads and shifting consumption to off-peak periods is delayed, if not fully displaced.

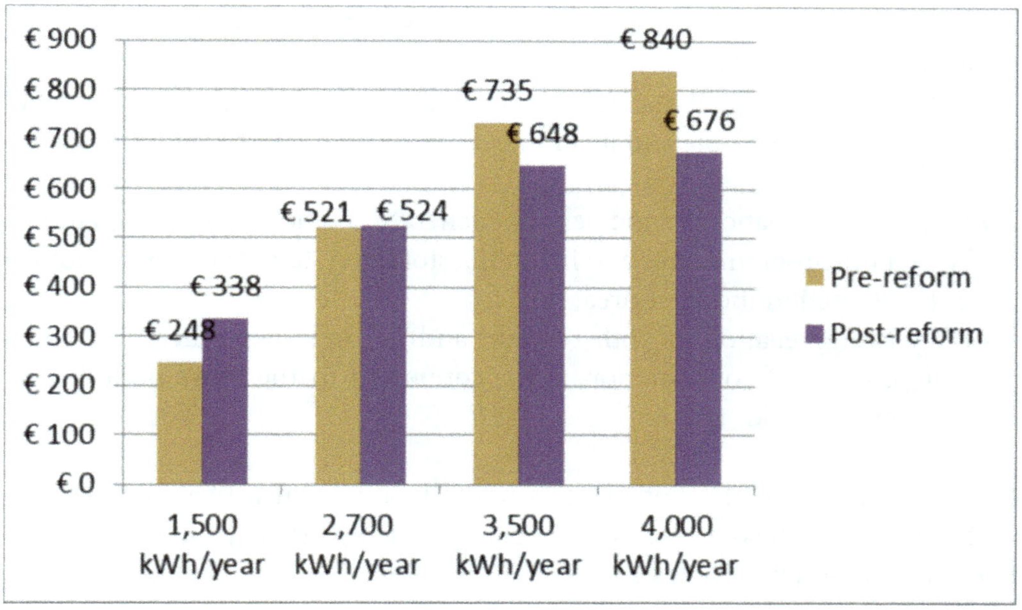

Figure 17: Comparison of total costs of the electricity bill pre and post reform for 3 kW user.

Source: Energy and Strategy (2016).

Demand response and grid services from renewables and storage

Since 2004, eligible consumers have been allowed to bid on the demand side of the wholesale market, however, demand-side resources do not have access to the balancing market and are not allowed to provide ancillary services, which are only provided by generators with installed capacity above 10 MVA. With regard to balancing responsibilities, RES are charged for their imbalances since 2012 with a given tolerance, under a continuously changing regulation (ARERA 2012). This rule exposed the balancing market to expensive manipulations and the rules are again under revision.

Thanks to the fact that Italy was the first country in Europe that adopted smart meters on a large scale, with more than 95 per cent of low-voltage consumers currently equipped with this technology (Meeus & Saguan 2011), the use of demand resources to manage the grid is feasible. Since 2010, a time-of-use tariff (peak and off-peak) has been mandatory for consumers in the captive market, and retailers offer time-of-use prices to all types of consumers; in spite of this only consumers with contracted capacity higher than 55 kW are charged on an hourly basis.

The Italian NRA, with Resolution no. 300/2017/R/eel (ARERA 2017), has started a process to open the ancillary services market to non-relevant and non-programmable generation units, as well as to consumption units and to storage systems. To this aim, such installation can be aggregated in virtual units (UVA, which stands for Unità Virtuali Abilitate – Virtual Enabled Units) and take part in the market as a single aggregated participant. In the abovementioned Resolution, four types of virtual units are addressed:

- UVA-C, aggregation of consumption units situated in the same area (areas are defined by Terna);
- UVA-P, aggregation of non-relevant generation units (either programmable or non-programmable and including storage systems) situated in the same area;
- UVA-M, aggregation of non-relevant generation units (either programmable or non-programmable and including storage systems) and consumption units situated in the same area;
- UVA-N, aggregation of both relevant and non-relevant generation units, and possibly of consumption units, connected to the same node of the transmission grid.

Currently (June 2019), pilot projects have already been activated and these units have already taken part in the ancillary services market, but it is too soon to evaluate their effectiveness.

The case of small islands

The Italian coastal areas are characterised by several small islands, which are not connected to the mainland electricity network. While they can represent an opportunity to develop local systems entirely reliant upon low-carbon sources, most of these islands are supplied by fossil-fuel generators, largely oversized compared to winter demand, in order to cover summer peak loads (Smart Island 2016). The decarbonisation of these local systems has never been undertaken because consumers on these islands have been subsidised by the rest of final consumers (€60 million/year) (ARERA 2014a): they pay the same tariffs as the rest of the country, plus local operators can recover costs in entirety, leading to significant profit and opposition to establishing more innovative solutions.

Governance drivers for distributed energy resources

A regulation-driven model for decentralisation: SEU

The business model created by the Italian regulatory framework for distributed generation, Sistemi Efficienti di Utenza (SEU), is affected by significant limitations. A SEU is a system where one or more plants (RES or high-efficiency CHP plants) managed by the same producer (which could also differ from the end user) are directly connected through a private connection (with no obligation to connect third parties) with one final consumer and all the elements of the SEU (plants, consumption site, connection, network) are included in an area available to the final consumer itself (ARERA 2013). The regulation for SEU, which has already been affected by retroactive changes, has so far mainly supported PV projects with an installed capacity lower than 20 kW_p (GSE 2016), and represents a significant barrier for the development of the scheme with regard to the one-to-one restriction and the ownership of the whole area where the plant is installed.

In January 2019, Italy submitted to the European Commission its draft proposal of the National Energy and Climate Plan (NECP) (Ministero dello Sviluppo Economico, Ministero dell'Ambiente, Ministero delle Infrastrutture 2018). The Plan is aimed to accelerate decarbonisation, to enhance the energy decentralisation and to ensure security of supply, while promoting energy efficiency and the electrification of consumption. Among the main goals of the Plant there is the increase of the quota of energy generated by RES up to 30% of the national gross energy consumption and a 33% reduction of greenhouse gas for non-ETS sectors emissions.

The NECP also declares the intention to proceed with the transposition of the Directive 2018/2001 (Directive 2018/2001), with particular reference to individual and collective self-consumption initiatives (energy communities). To enhance their development, the main instrument should be the application of network and system charges only on the electricity which is procured from the network, while the electricity which is self-consumed should not be burdened by such charges. The transposition is still in process at the time of writing (June 2019), but it is likely that it will enhance the quota of self-consumption, with particular reference to renewable generation units: 80% of self-consumed energy is currently produced by gas-fired cogeneration plants (GSE 2018).

Other support policies for decentralisation

Various support measures are available, but they are not conceived in the perspective of creating a full, decentralised energy system.

In 2004 the Italian legislator introduced a white certificates system (D.M. 20 July 2004) with an obligation placed on electricity and gas distributors (with a threshold of 50,000 consumers). Distributors are allowed to invest in energy efficiency initiatives themselves or to purchase the certificates from ESCOs that undertake investments in this field.

Tax deductions (Legge 296/2006) of up to 65 per cent of the investment cost over a ten-year time span are available for energy efficiency measures such as solar heating collectors, condensing boilers, high-efficient heat pumps and biomass boilers. Such deductions have been the key driver for energy efficiency improvements in the building sector in Italy, with more than 14.2 million interventions from 1998 to 2016 (Servizio Studi Camera dei Deputati 2016).

Further incentive for energy efficiency initiatives and renewable thermal energy is provided by 'Conto Termico' (D.M. 28 December 2012; D.L. 91/2014), a contribution available for public authorities and households that covers part of the costs incurred and is paid off in annual instalments, from 2 to 5 years.

Storage tests and smart grid tests

Many European and national programmes provided funding for storage facilities and smart grid solutions, in order to facilitate the integration of RES and establish new modalities for the operation of transmission and distribution grids; one of the largest project in Italy financed the refurbishment of 1,605 km of transmission and distribution lines in Southern Regions (Ministero dello Sviluppo Economico 2014). Moreover, several smart grid projects were also implemented by the incumbents themselves, thanks to the possibility to include these expenses in the cost-recovery mechanism: Enel Distribuzione in 2015 committed itself for a total amount of € 343 million under different financial agreements (Mori 2015). Terna installed 35 MW of storage units in southern regions to increase the flexibility of the system and to absorb excess power from non-dispatchable RES in off-peak hours (Terna 2016).

NRA defined the mechanisms for the selective promotion of investments towards smart distribution systems in areas with a large penetration of DG (ARERA 2015a). These projects are related to new anti-islanding protection schemes, and real-time operation, monitoring and control strategies. However, the NRA itself encountered barriers in carrying out these projects, and the most significant one has been the lack of involvement and participation of active users. Even if the distributor covered all the costs, some users have rejected the experiment because they lacked direct immediate benefits and were scared by the problems that could occur during the process.

Relying on conventional network operators to test smart grid solutions is preferable in terms of the stability of the system, but this solution is risky with respect to the exclusion of other operators that could supply more innovative technologies to regulate DG.

2.7.5 Conclusions

Italy complied with the European decarbonisation targets, even if the regulation of the electricity system was not able to promote a proper coherent governance for the transition towards a decentralised paradigm. As a matter of fact:

- The reform of the household's tariff is not in favour of efficient consumption units and does not support local generators; moreover, the expected results in terms of electrification of final uses are not guaranteed;
- The restrictions on aggregation and on the provision of services from DG, as well as the barriers to coupling storage units and generators, hamper the adoption of innovative technologies; in this perspective, the adoption of Virtual Enabled Units can represent a significant innovation, which however is still at its early stage;
- In principle the authors agree with the opportunity to fully implement the liberalisation process, but considering that significant informative asymmetries are still available in the market, an aggregator with no-profit target like the single buyer still represents a useful benchmark for the market;
- The prohibition to sell electricity from local plants to adjacent entities is preventing the establishment of decentralised supply, but the commitment for the transposition of the Directive 2018/2001, with particular reference

to energy communities and to jointly acting renewable self-consumers can be a driver for decentralisation.

In spite of the latest improvements, these policies have resulted in the protection of the rent of incumbents. Conversely, the low-cost technologies available allow sustainable facilities and demand aggregators to provide network services; the integration of thermal and electric loads could also facilitate the management of intermittent electricity sources if efficient solutions are in place.

Driving the change in the energy sector requires great regulatory vision and the ability to balance the needs of the incumbents with the opportunities opened up by the innovative options. The flattening electricity demand and the shrinking prices do not make it easy to put in place novel instruments, but the new market design should fit with low-carbon and decentralised energy sources.

2.7.6 References

ARERA. (2007), Delibera n. 280/07. Modalità e condizioni tecnico-economiche per il ritiro dell'energia elettrica ai sensi dell'articolo 13, commi 3 e 4, del decreto legislativo 29 dicembre 2003, n. 387/03, e del comma 41 della legge 23 agosto 2004, n. 239/04. Milan: Autorità di Regolazione per Energia Reti e Ambiente (ARERA)

ARERA. (2012), Deliberazione 20 dicembre 2012 570/2012/R/EFR. Testo integrato delle modalità e delle condizioni tecnico-economiche per l'erogazione del servizio di scambio sul posto: condizioni per l'anno 2013. Milan: Autorità di Regolazione per Energia Reti e Ambiente (ARERA)

ARERA. (2013), Deliberazione 12 dicembre 2013 578/2013/R/EEL. Regolazione dei servizi di connessione, misura, trasmissione, distribuzione, dispacciamento e vendita nel caso di Sistemi Semplici di Produzione e Consumo. Milan: Autorità di Regolazione per Energia Reti e Ambiente (ARERA)

ARERA. (2014a), Memoria 3 luglio 2014 332/2014/I/EEL. Conversione in legge del Decreto Legge 24 Giugno 2014, n. 91. Milan: Autorità di Regolazione per Energia Reti e Ambiente (ARERA)

ARERA. (2015a), Documento per la consultazione DCO 255/2015/R/EEL. Smart Distribution System: promozione selettiva degli investimenti nei sistemi innovativi di distribuzione di energia elettrica. Milan: Autorità di Regolazione per Energia Reti e Ambiente (ARERA)

ARERA. (2015b), Documento per la consultazione DCO 34/2015/R/EEL. Riforma delle tariffe di rete e delle componenti tariffarie a copertura degli oneri generali di sistema per i clienti domestici di energia elettrica. Milan: Autorità di Regolazione per Energia Reti e Ambiente (ARERA)

ARERA. (2015c), Delibera 4 giugno 2015, 271/2015/R/com. Avvio di procedimento per la definizione del percorso di riforma dei meccanismi di mercato per la tutela di prezzo dei clienti domestici e delle piccole imprese nei settori dell'energia elettrica e del gas naturale- tutela 2.0. Milan: Autorità di Regolazione per Energia Reti e Ambiente (ARERA)

ARERA. (2015d), Delibera 296/2015/R/com. Disposizioni in merito agli obblighi di separazione funzionale (unbundling) per i settori dell'energia elettrica e del gas. Milan: Autorità di Regolazione per Energia Reti e Ambiente (ARERA)

ARERA. (2015f), Deliberazione 10 marzo 2015 95/2015/I/eel. Proposta al Ministro dello Sviluppo Economico per l'anticipazione della fase di piena attuazione del mercato della capacità. Milan: Autorità di Regolazione per Energia Reti e Ambiente (ARERA)

ARERA. (2016), Relazione 339/2016/I/EFR. Stato di utilizzo e di integrazione degli impianti di produzione alimentati dalle fonti rinnovabili e degli impianti di cogenerazione ad alto rendimento. Relazione sullo stato dei servizi. Milan: Autorità di Regolazione per Energia Reti e Ambiente (ARERA)

ARERA. (2017), Delibera 5 Maggio 2017, 300/2017/R/EEL. Prima apertura del Mercato per il Servizio di Dispacciamento (MSD) alla Domanda Elettrica ed alle Unità di Produzione anche da fonti rinnovabili, non già abilitate, nonché ai sistemi di accumulo. Istituzione di progetti pilota in vista della costituzione del Testo Integrato Dispacciamento Elettrico (TIDE) coerente con il Balancing Code Europeo. Milan: Autorità di Regolazione per Energia Reti e Ambiente (ARERA)

ARERA. (2018), Relazione annuale sullo stato dei servizi e sull'attività svolta. Milan: Autorità di Regolazione per Energia Reti e Ambiente (ARERA)

Biancardi, A. (2016), Audizione di rappresentanti dell'Autorità per l'Energia Elettrica, il Gas e il Sistema Idrico. Commissione Industria, Commercio, Turismo del Senato. Resoconto n. 255 del 13/07/2016.

Clò, A. (2014), L'impervio e incompiuto cammino verso il mercato unico europeo dell'energia, in Riforme elettriche tra efficienza ed equità, 1st ed., A. Clò, S. Clò, F. Boffa, Ed. il Mulino, Bologna.

Clò, S., Cataldi, A. & Zoppoli, P. (2015), The merit-order effect in the italian power market: the impact of solar and wind generation on national wholesale electricity prices. *Energy Policy, 77*, 79–88.

Decreto Legislativo 16 marzo 1999, n. 79. Attuazione della direttiva 96/92/CEE recante norme comuni per il mercato interno dell'energia elettrica.

Decreto Legislativo 29 dicembre 2003 n. 387. Attuazione della direttiva 2001/77/Ce relativa alla promozione dell'energia elettrica prodotta da fonti energetiche rinnovabili nel mercato interno dell'elettricità, Gazzetta Ufficiale della Repubblica Italiana, del 31 gennaio 2004 n. 25.

Decreto del Ministro delle attività produttive di concerto con il Ministro dell'ambiente e della tutela del territorio e del mare del 28 luglio 2005. Criteri per l'incentivazione della produzione di energia elettrica mediante conversione fotovoltaica della fonte solare.

Decreto Legislativo 8 febbraio 2007, n.20. Attuazione della direttiva 2004/8/CE sulla promozione della cogenerazione basata su una domanda di calore utile nel mercato interno dell'energia, nonche' modifica alla direttiva 92/42/CEE. **Decreto del Presidente** del Consiglio dei Ministri, 11 maggio 2004. Criteri, modalità e condizioni per l'unificazione della proprietà e della gestione della rete elettrica nazionale di trasmissione.

Decreto Legislativo n. 102/2014. Attuazione della direttiva 2012/27/UE sull'efficienza energetica, che modifica le direttive 2009/125/CE e 2010/30/UE e abroga le direttive 2004/8/CE e 2006/32/CE.

Decreto Legge 24 giugno 2014, n. 91. Disposizioni urgenti per il settore agricolo, la tutela ambientale e l'efficientamento energetico dell'edilizia scolastica e universitaria, il rilancio e lo sviluppo delle imprese, il contenimento dei costi gravanti sulle tariffe elettriche, nonche' per la definizione immediata di adempimenti derivanti dalla normativa europea.

Decreto Legge 30 dicembre 2016 n. 244. Proroga e definizione di termini. Decreto Legge convertito con modificazioni dalla Legge 27 febbraio 2017, n. 19.

Decreto Ministeriale. (2007), Decreto Ministeriale 19 febbraio 2007, Criteri e modalita' per incentivare la produzione di energia elettrica mediante conversione fotovoltaica della fonte solare, in attuazione dell'articolo 7 del decreto legislativo 29 dicembre 2003, n. 387.

Decreto Ministeriale. (2010), Decreto Ministeriale 6 agosto 2010. Incentivazione della produzione di energia elettrica mediante conversione fotovoltaica della fonte solare.

Decreto Ministeriale. (2012), Decreto Ministeriale 5 luglio 2012. Attuazione dell'art. 25 del decreto legislativo 3 marzo 2011, n. 28, recante incentivazione della produzione di energia elettrica da impianti solari fotovoltaici (c.d. Quinto Conto Energia).

Decreto Ministeriale. (2012), Decreto Ministeriale 28 dicembre 2012. Incentivazione della produzione di energia termica da fonti rinnovabili ed interventi di efficienza energetica di piccole dimensioni.

Decreto Ministeriale. (2014), Decreto Ministeriale 30 giugno 2014 – Disciplina del mercato della capacità.

Directive 1996/92/EC of the European Parliament and of the Council of 19 December 1996 concerning common rules for the internal market in electricity. Brussels: European Parliament

Directive 2001/77/EC of the European Parliament and of the Council of 27 September 2001 on the promotion of electricity produced from renewable energy sources in the internal electricity market. Brussels: European Parliament

Directive 2003/54/EC of the European Parliament and the Council of 26 June 2003 concerning common rules for the internal market in electricity and repealing Directive 96/92/EC. Brussels: European Parliament

Directive 2009/28/EC of the European Parliament and the Council of 23 April 2009 on the promotion of the use of energy from renewable sources and amending and subsequently repealing Directives 2001/77/EC and 2003/30/EC. Brussels: European Parliament

Directive 2009/72/EC of the European Parliament and of the Council of 13 July 2009 concerning common rules for the internal market in electricity and repealing Directive 2003/54/EC. Brussels: European Parliament

Directive 2012/27/EU of the European Parliament and of the Council of 25 October 2012 on energy efficiency, amending Directives 2009/125/EC and 2010/30/EU and repealing Directives 2004/8/EC and 2006/32/EC. Brussels: European Parliament

Directive 2018/2001 of the European Parliament and of the Council of 11 December 2018 on the promotion of the use of energy from renewable sources (recast). Brussels: European Parliament

Disegno di Legge Costituzionale S. 1429-D. (2016), Disposizioni per il superamento del bicameralismo paritario, la riduzione del numero dei parlamentari, il contenimento dei costi di funzionamento delle istituzioni, la soppressione del CNEL e la revisione del titolo V della parte II della Costituzione" (approvato, in seconda deliberazione, dal Senato con la maggioranza assoluta dei suoi componenti, già approvato, in prima deliberazione, dal Senato, modificato, in prima deliberazione, dalla Camera, ulteriormente modificato, in prima deliberazione, dal Senato e approvato, senza modificazioni, in prima deliberazione, dalla Camera).

ENEA. (2016a), Analisi trimestrale del sistema energetico italiano. Rome: Agenzia nazionale per le nuove tecnologie, l'energia e lo sviluppo economico sostenibile (ENEA).

ENEA. (2016b), Parigi e oltre. Gli impegni nazionali sul cambiamento climatico al 2030. Rome: Agenzia nazionale per le nuove tecnologie, l'energia e lo sviluppo economico sostenibile (ENEA).

Energy and Strategy. (2016), Energy efficiency report. Milan: Energy & Strategy Group.

European Commission. (2014), Communication from the Commission to the European Parliament, the Council, the European Economic and Social Committee and the Committee of the Regions. A policy framework for climate and energy in the period from 2020 to 2030.

Federconsumatori. (2014), Politiche, caro bollette, mercato energia: quail risposte per I consumatori?

Font, N. (2002), La politica ambientale, in S. Fabbrini, F. Morata, L'Unione Europea. Le politiche pubbliche, Roma-Bari, Laterza, 2002.

GSE. (2016) Rapporto Attività 2015.

GSE. (2018). Audizione X Commissione Senato, 18 Settembre 2018.

GSE. (2019a), Rapporto Attività 2018. Rome: Gestore Servizi Energetico (GSE).

GSE. (2019b), Rapporto Statistico 2017. Rome: Gestore Servizi Energetico (GSE).

IEA. (2016), Re-powering markets. Market design and regulation during the transition to low-carbon power systems. Paris: International Energy Agency.

Kuzemko, C., Lockwood, M., Mitchell, C. & Hoggett, R. (2016), Governing for sustainable energy system change: Politics, contexts and contingency. *Energy Research and Social Science, 12*, 96–105.

Legge n. 9, 9 gennaio 1991. Norme per l'attuazione del nuovo piano energetico nazionale: aspetti istituzionali, centrali idroelettriche ed elettrodotti, idrocarburi e geotermia, autoproduzione e disposizioni fiscali.

Legge n. 481, 14 novembre 1995. Norme per la concorrenza e la regolazione dei servizi di pubblica utilita'. Istituzione delle Autorita' di regolazione dei servizi di pubblica utilita'.

Legge 27 ottobre 2003, n. 290. Conversione in legge, con modificazioni, del decreto-legge 29 agosto 2003, n. 239, recante disposizioni urgenti per la sicurezza del sistema elettrico nazionale e per il recupero di potenza di energia elettrica. Deleghe al Governo in materia di remunerazione della capacità produttiva di energia elettrica e di espropriazione per pubblica utilità.

Legge n.296, 27 dicembre 2006. Disposizioni per la formazione del bilancio annuale e pluriennale dello Stato (legge finanziaria 2007), in Gazzetta Ufficiale del 27 dicembre 2006 n. 299.

Legge n.129, 13 agosto 2010. Conversione in legge, con modificazioni, del decreto-legge 8 luglio 2010, n. 105, recante misure urgenti in materia di energia. Proroga di termine per l'esercizio di delega legislativa in materia di riordino del sistema degli incentivi.

Legge Costituzionale 18 ottobre 2001, n. 3. Modifiche al titolo V della parte seconda della Costituzione.

Lorenzoni, A. (2014), Nel mercato futuro dell'energia è problematico lo spazio per le "vecchie" utility, see http://www.qualenergia.it/articoli/20140529-nel-mercato-futuro-energia-problematico-lospazio-per-le-utility.

Lorenzoni, A. (2015), Come il fotovoltaico italiano sta mostrando la strada del cambiamento, see http://www.qualenergia.it/articoli/20150323-come-il-fotovoltaico-italiano-sta-mostrando-la-strada-del-cambiamento.

Meeus, L. & Saguan, M. (2011), Innovating grid regulation to regulate grid innovation: From the Orkney Isles to Kriegers Fliak via Italy. *Renewable Energy, 36*, 1761–1765.

Ministero dello Sviluppo Economico. (2014), Rinnovabili ed efficienza energetica. Un racconto lungo una programmazione. I fondi europei 2007-2013 un'opportunità di sviluppo per il Sud.

Ministero dello Sviluppo Economico. (2018), La situazione energetica nazionale nel 2017.

Ministero dello Sviluppo Economico, Ministero dell'Ambiente e della Tutela del Territorio e del Mare, Ministero delle Infrastrutture e dei Trasporti. Proposta di Piano Nazionale Integrato per l'Energia e il Clima. 31/12/2018.

Mori, S. (2015). Audizione presso X e XIII Commissioni del Senato della Repubblica. 6 maggio 2015.

Nextville. (2013), Vademecum Nextville 2013. Fonti rinnovabili. Milan: Edizioni Ambiente.

Polo, M., Cervigni, G., D'Arcangelo, F.M. & Pontoni F. (2014), La regolazione delle reti elettriche in Italia. IEFE Research Report no. 15 - June 2014.

Ranci, P. (2014), La liberalizzazione del mercato elettrico al dettaglio tra efficienza ed equità sociale, in Riforme elettriche tra efficienza ed equità, 1st ed., A. Clò, S. Clò, F. Boffa, Ed. il Mulino, Bologna.

REN21. (2016), Renewables 2016 global status report. Retrieved February 10, 2019, from www.ren21.net/wp-content/uploads/2016/05/GSR_2016_Full_Report_lowres.pdf

Ricerca sul Sistema Energetico (RSE). (2015), Energia elettrica, anatomia dei costi. Milan: Editrice Alkes.

Servizio Studi Camera dei Deputati. (2016), Il recupero e la riqualificazione energetica del patrimonio edilizio: una stima dell'impatto delle misure di incentivazione. 9 September 2016.

Smart Island. (2016), http://www.smartisland.eu/

Terna. (2016), https://www.terna.it/it-it/azienda/chisiamo/ternastorage.aspx.

UNFCCC. (2015), Adoption of the Paris Agreement, 12 December 2015.

Unione Petrolifera (2016). Data Book 2016. Energia e Petrolio.

2.8 California versus New York: policy implementation via Investor-Owned Utilities or Distribution System Provider?

Catherine Mitchell[28]

2.8.1 Introduction

Energy systems are changing all over the world. In the United States of America (USA), the different states follow very different energy policies. Some are at the forefront of global energy policy thinking, including California (CA) and New York state (NYS). However, CA and NYS, whilst both progressive in terms of greenhouse gas (GHG) reduction policies, have very different principles underlying their energy system governance[29] and therefore implement their energy policies in very different ways. CA is one of the world's earliest movers in terms of energy system transformation and, on most metrics, outperforms NYS. However, NYS in 2014 put in place the New York Reforming the Energy Vision (NY REV), and although it is too early to judge how successful it is, it has introduced new ideas and arguments about the necessary governance constituents for energy system transformation, and is acting as a first mover in certain significant ways.

[28] Work undertaken as a result of the UK EPSRC IGov Research award http://projects.exeter.ac.uk/igov/; and thanks to Carl Linvill of RAP, and Rudi Stegemoeller, ex of RAP and now NY PSC, for their conversations.

[29] Governance is thought of as the combination of policies, institutions, regulations, market and network rules and incentives, and the process by which the governance design details (i.e. the details of a RE policy, or a market rule) are agreed.

How to cite this book chapter:

Mitchell, C. 2020. California versus New York: policy implementation via Investor-Owned Utilities or Distribution System Provider?. In: Burger, C., Froggatt, A., Mitchell, C. and Weinmann, J. (eds.) *Decentralised Energy — a Global Game Changer.* Pp. 133–156. London: Ubiquity Press. DOI: https://doi.org/10.5334/bcf.i.

This country report endeavours to capture those differences and similarities between CA and NYS. In many ways, the heart of a comparison between CA and NY is:

- whether the more cautious governance approach of CA with respect to markets and the ways it places responsibility via regulation on its utilities to execute its policies, often through procurement, is working as well, or better, than the NYS's avowedly new 'balance' approach between regulation and markets of the Reforming the Energy Vision (REV);
- whether, at the end of the day, there is not a great deal of practical difference in outcomes between the two in terms of practical innovation and change, even if their avowed governance principles might suggest there is, and this therefore implies their differences do not really matter.

CA is far ahead of NYS in some ways in terms of renewable energy deployment. However, the changes within the global energy system are increasingly clustering around decentralisation of technologies; the move to decarbonisation; the inclusion of digitalisation within energy system operation and market/platform transactions; and the increasing preference for some degree of democratisation of the energy system via customer choice, new ownership or involvement of new stakeholders or investors. Together, these four 'Ds' are known as D4.

At this point in time, NYS's intended energy pathway might appear to be better suited to these changes (because of its efforts to encourage new entrants; and new, dynamic ways of system operation and valuation of DER; and customer choice) and therefore may – relatively – better benefit the citizens of NYS than CA in terms of energy system transformation over the longer term. On the other hand, NYS may find it just too hard to push through fundamental changes across the spectrum of energy system operation and markets. Only time will tell whether CA or NYS is the more successful governance model for energy system transformation.

Notwithstanding this, the chapter is arguing, at root, that any comparison between them has to be put in context. CA has been supporting sustainable energy since the 1970s. What NYS was able to do in 2014 was 'start afresh' and has enunciated a new approach to energy system regulation to enable a new, more dynamic system operation, with a new value proposition in keeping with D4. There are multiple documents which describe, or predict, a very different global energy system by 2030 (for example see Navigant 2017). However, there are very few country energy systems around the world which are 'walking the walk' and changing rules and incentives so that the value of D4 is available and accessible for new actors and new ways of doing things. This is why both CA and NYS are so interesting.

For any other US state or country which is trying to work out what governance system would most suit them for energy system transformation, valuable lessons can be learned from both CA and NYS.

Overall though, the chapter argues that the underlying principles of the NYS regulatory reform are potentially transformational because it is trying to create a new 'value' proposition within energy system governance [1] by arguing to move the 'heart' of the energy system to the distribution level; [2] to create distribution system providers to facilitate and coordinate markets at that level, and then to 'nest' up that local market up to the wholesale level enabling a dynamic valuation of distributed energy resources; and [3] to both confront, and provide solutions, to the altered and changing nature of energy system provision. The NY REV is illuminating the fundamental issues (including the difficulties) which need to be addressed when undertaking energy system transformation, and in this way the efforts of the NY REV can only be helpful to wider global energy system transformation practice.

Countries may enact this new value proposition differently from NYS, but the ideas that the NY REV have unleashed can be expected to roll out around the world. This chapter, however, also argues that the governance principles or Vision of the NY REV should be viewed distinctly from what is happening in NYS on the ground. The latter, so far, is more about getting information, processes, methodologies and value of DER sorted out rather than delivering much practical change. These are the vital building blocks for that transformational change, and even if NYS runs into difficulties – as a first mover, it is acting as a laboratory for the rest of the world. However, all countries and States have different cultures, geographies and energy system history. Simply transferring NY REV regulation to a country cannot be expected to work. What can be expected to be useful is to understand what energy system issues the NY REV principles and processes were chosen to address (and why), and then to assess whether those principles and processes could be of help in other countries.

This chapter first (very briefly) reviews the CA energy policies, regulation and ethos; then NY's energy regulation and ethos; and then provides the comparison. This chapter is not attempting a complete overview and comparison of the CA and NYS energy policies. It is attempting to highlight the key characteristics and pieces of legislation of both places for DER governance. For those who would like to have a more detailed overview of how the United States regulates its energy industry, please see.[30] The United States is very different from, say, Europe because 'utilities' are often combined distribution and supply, and often with a default tariff and limited, or no, competitive retail at the domestic level. Notwithstanding these differences, there is still a lot that the rest of the world can learn in terms of DER governance ideas.

---

30 http://www.raponline.org/knowledge-center/electricity-regulation-in-the-us-a-guide-2/.

2.8.2 California

Overview of Californian energy policy

CA has been at the forefront of global energy policy since the 1970s. Their reaction to the oil crises and energy insecurity of the early 1970s was to put in place policies to support energy efficiency measures and renewables.[31] The Public Utility Regulatory Policies Act (PURPA) of 1978 was one of the first countries/US states in the world to legislate to support renewable energy implementation.[32] CA is an example of a state/country which has had long-term, political commitment and leadership with respect to sustainable energy, via both Republican and Democrat Governors. One result of this early mover advantage has been that the air quality aspects of energy use are also an important and integrating focus of CA energy policy. Another result is that policy is driven by the CA legislature and the various agencies – of which CA has many – are then responsible for implementing the state policy. It is therefore a top-down system.

The main CA implementation agencies are the CEC (the California Energy Commission); the CPUC (the California Public Utilities Commission); CAISO (the Californian Independent System Operator); and CARB (the Californian Air Resources Board). There has been criticism in the past that those agencies do not work well together.[33] However, in November 2016, the California Public Utility Commission (CPUC) published a 7 page Distributed Energy Resources (DER) Action plan – intended to provide a Vision for the way forward for energy within CA, and as part of that a way to integrate the different energy institutions and their activities (California State 2016).

Energy and climate trends in California

GHG emissions in CA have been reasonably similar between 1990 and 2014. Senate Bill (SB) 32 (discussed below) is intended to bring them down significantly by 2030 (California State 2006). However, it is the transportation and electricity sectors which have managed to reduce emissions[34] between 2000 and 2014, whilst industrial, commercial and residential, agriculture, recycling and waste have more or less stayed the same. The provision of electricity generation capacity has increased significantly since 2001 (around 56,000 MWs) to

[31] During the 70s, a new nuclear program was considered but once PURPA was in place this receded.

[32] The other country, at that time, which reacted in a similar way to the 1970s oil crises, and which also set about supporting renewables, was Denmark. Both those countries have gone on to become world sustainable technology leaders.

[33] For a rather old analysis of this see http://www.hoover.org/sites/default/files/research/docs/127600810-renewable-and-distributed-power-in-california.pdf.

[34] http://docketpublic.energy.ca.gov/PublicDocuments/16-IEPR-01/TN215418_20170118T122654_Proposed_Final_2016_Integrated_Energy_Policy_Report_Update_Clea.pdf.

around 80,000 MWs in 2014, with natural gas and renewables making up the majority of the new capacity.[35] Renewable electricity now provides about 30 per cent of retail electricity sales, which includes electricity imported from other states (CEC 2018). Solar has increased from 6800 MW in 2001 to 16,200 MW in 2017. Out of a total of 27,800 MW of renewables.[36]Moreover, California is a leader for implementation across technologies for the United States. 'California represents 49 per cent of all the distributed solar that's been installed; it represents 49 per cent of all the distributed storage that's been built; and it represents 47 per cent of all the plug-in electric vehicles in the United States (Wesoff 2017).

The pillars of Californian energy policy

Within this pro-environment context, CA energy policy had a major existential crisis in 2001 which has had a profound effect on subsequent energy policy, and their attitude to markets. CA implemented a privatisation of their electricity system in 2001 – which failed (Sweeney 2002).[37] Since then, CA has been extremely cautious about introducing more avowedly 'market' based policies or institutional reforms to the CA energy system.

Broadly, CA has three main investor owned utilities (IOUs) – Pacific Gas and Electric, San Diego Gas and Electric, and Southern California Edison which are joint distribution and supply companies which supply about 70 per cent of electricity. The other 30 per cent is supplied by Municipal-Owned Utilities or 'munis', with the biggest being Sacramento Municipal Utility District and the Los Angeles Department of Water and Power. The IOUs buy electricity from wholesale markets; and are the main executors of CA state energy policy by procuring renewables, providing contracts for energy efficiency measures, demand response, storage and so on, as they are mandated to do. They then sell to customers in what was their monopoly areas, via distribution grids that they own. This institutional set up has also meant that the problems and solutions of energy system transformation – such as working out how to pay for networks with increasing amounts of onsite generation; how to fulfil state energy policy goals, such as energy efficiency programmes; how to integrate rate design (or

[35] http://docketpublic.energy.ca.gov/PublicDocuments/16-IEPR-01/TN215418_20170118T122654_Proposed_Final_2016_Integrated_Energy_Policy_Report_Update_Clea.pdf.

[36] http://www.energy.ca.gov/renewables/tracking_progress/documents/renewable.pdf.

[37] See James L Sweeney, The California Electricity Crisis (2002) Hoover Press, or for a condensed version: http://web.stanford.edu/~jsweeney/paper/Lessons percent20for percent20the percent20Future.pdf; or for a very quick overview http://projects.exeter.ac.uk/igov/lessons-from-america-worrying-analogies-between-the-emr-process-and-the-california-electricity-crisis-2001/.

tariff structures as they are often known in other jurisdictions) has been implemented via regulation and the IOUs.

Technically, there is retail competition although in a de facto sense competition mainly occurs in the non-domestic sectors. It is possible to have Community Choice Aggregation (CCA) which is when a community (and therefore domestic customers) is able, under certain rules, to procure energy for the community (CPUC 2012); and for Electric Service Providers (ESP, the equivalent of the European supplier concept) that offer electric services directly to retail (including domestic) customers within the main IOU service areas. Some CCAs can be large, although they are still reliant on their IOU or muni for billings and so on (Trabish 2017). In practice, ESPs do not provide energy to domestic customers. As a result, a domestic customer in CA who does not live in a CCA area can de facto only buy their electricity from the IOU which works in their area. CA is therefore a state with limited domestic retail competition. There is competition in the IOU take-up of resources to the extent that the utility either buys via a wholesale market or procures resources (as a result of regulatory requirements) based on competitive bidding.[38] The utilities have therefore become the executor of the state energy policies.

In 2003, the California Energy Commission (CEC) established the California Statutory Energy Loading Order via the CA Energy Action Plan, 2003, and then updated it in 2008 (CEC 2008). This Loading Order required IOUs to procure energy efficiency and demand response ahead of all other resources, including ahead of priority access for renewables.

In parallel to the CEC's Loading Order, 'modern' Californian energy policy was founded in State Assembly Bill (AB) 32 (California State 2018) – the Global Warming Solutions Act, 2006 – which formally brought air quality and climate change together along with security and affordability issues. The 2006 bill has been strengthened over time by Senate Bill (SB) 350[39] in 2013; SB 32 in 2016 (which has a target of 40 per cent reduction of GHG from 1990 levels by 2030); and AB 197 in 2016 which aims to ensure that the State's implementation of these policies is transparent and equitable, and that their benefits also flow to the disadvantaged (for an overview of state papers please see[40]).

An Integrated Energy Policy Report (IEPR) is published every 2 years by the California Energy Commission (see CEC 2018). The IEPR reports attempt to provide an overview, and explain the inter-relationships, of the CPUC, CAISO, CARB and CEC policy measures. But it has been the CPUC's 7 page Distributed Energy Resources Action Plan (CPUC 2016), published in November 2016, which has set out a time plan for Actions to integrate, and take forward, all the various measures within California and its Agencies with respect to DER.

---

[38] For a rather old analysis of this see http://www.hoover.org/sites/default/files/research/docs/127600810-renewable-and-distributed-power-in-california.pdf.

[39] http://www.energy.ca.gov/sb350/.

[40] http://www.cpuc.ca.gov/sb350/.

Arguably, because of this, CA energy policy has five key pillars, and it is important to note the different agencies involved in these different pillars:

- Electricity system decarbonisation through renewables implementation and energy efficiency policies (RETI 1.0,[41] RETI 2.0,[42] and Long Term Procurement Planning) via the CEC.
- Decarbonisation through strengthening regional markets and creating a Regional Independent System Operator (RSO) and Energy Imbalance Market (EIM) via CAISO.
- Decarbonisation by building distributed energy resources (DG, EE, DR, storage, and EVs) via the CPUC.
- Optimising decarbonisation across sectors (electricity, building and transportation) via Integrated Resource Planning, coordinated by the CPUC.
- Valuing carbon, and air quality, policies (including the cap and trade scheme) via CARB.

This section focuses on the decarbonisation by building distributed energy resources (DG, EE, DR, storage, and EVs) via the CPUC, since it is most analogous to the NY REV Public Service Commission work. As in NYS, the CA state is providing support and a process to deliver renewables, energy efficiency and so on. There is a clear expectation that the Regulator (the CPUC in CA and the PSC in NYS) delivers regulatory mechanisms which complement policies.

Decarbonisation by building DER

The CPUC Scoping Note (CPUC 2014) of 2014 set up Distribution Resource Plan (DRP) proceedings (CPUC 2013) (the CA equivalent of the 2015 NYS Distribution System Implementation Plans) whereby CA IOU utilities are required to produce a DRP (DRPWG 2013). The Scoping Note argued that the underlying rationale for promoting increased deployment of DERs is that they have a critical role in meeting CA's GHG reduction policy. The goals of the DRP Plans were to:

- modernise the electric distribution system to accommodate two-way flows of energy and energy services throughout the IOU networks;
- enable customer choice of new technologies and services that reduce emissions and improve reliability in a cost efficient manner, but not to the extent that domestic customers can switch suppliers if not in a CCA area;
- to animate opportunities for DER to realise benefits through the provision of grid services, and to enable a plug and play distribution grid for DER (pp. 3–5).

---

[41] http://www.energy.ca.gov/reti/reti_1.html.

[42] http://www.energy.ca.gov/reti/.

The intent of the Scoping Note was very little different in terms of outcomes from that set out in the 2014 Vision Paper of the New York Reforming the Energy Vision (NYREV 2014) and discussed below. Moreover, the DRPs were also to be aligned with *More Than Smart* (RENSWICK Institute 2015), a progressive initiative which paints a much more integrated and market-based future and puts forward ideas about how to get there (CPUC 2014). Whilst this might imply a future with more domestic customer choice, CA has not clarified the point.

The DRPs have been followed up with the November 2016 CPUC DER Action Plan, as described above CPUC (2016), which has five Actions to be undertaken in 2017; three in 2018 and one in 2019. At root, the DER Action Plan process should work out:

- How to accommodate more DERs cost effectively, and work out the value of DERs when they are procured either individually (i.e. DG PV alone) or as a portfolio (DG PV plus storage etc).
- Make the wholesale and retail markets more responsive to DERs.
- Link DER more with CAISO, including with storage (CAISA 2014a; CAISA 2014b). CAISO has long had contracts for demand response and so forth but it now beginning to develop contracts with DER providers, or DERPs, which must register to (potentially) access a new revenue stream (CAISA 2018).
- Encourage a more integrated CA energy system.

Whilst the intentions are very similar to those in NYS, the means of assessing DER resource and value – as set out through the ICA and the LNBA) were very different in CA from NYS. The CA methodology was technical, economically static and set within the conventional institutional and market structure, and therefore the outcomes are very different (Brockaway 2017).

2.8.3 New York state

The New York reforming the energy vision

In April 2014, Governor Cuomo of New York kicked off the New York Reforming the Energy Vision (NY REV). This encompasses multiple dimensions of regulated administrated programs (such as support for renewable energy), regulatory reform and new institutions – all of which are intended to work together to create an enabling environment for a transition to a sustainable energy future for New York state (NYS 2018).

At part of this, the New York Public Service Commission (NY PSC, the Energy Regulator) initiated a regulatory reform aspect of the Reforming the Energy Vision (NYREV 2014) with a Regulatory Order in April 2014 from the Commissioners of the PSC to 'transform New York State's industry with the objective of creating market-based, sustainable products and services that drive an increasingly efficient, clean, reliable, and customer-orientated industry. A podcast (Lacy 2014), by the then-Chairperson of the NY PSC, Audrey Zibelman, described how Hurricane Sandy was a major driver in the NYS decision to articulate a new Vision for energy. Hurricane Sandy provided the appetite for change of the NYS people, as well as the PSC focus of providing customers with the services they want – which includes security, cost effectiveness and so on.

In brief the NY REV is:

- Envisaging a decentralised energy system, with a new 'heart' at the distribution level, which is coordinated in a new way via a distribution system provider (DSP) with more values for more services via transactive energy markets/platforms.
- Envisaging a 'new' regulatory framework and basis, with more performance based regulation, more suited to meeting the challenges faced by the NY energy system, and where appropriate, incentivising the solutions rather than regulating for certain outcomes.
- Envisages services aimed at fulfilling individualised customer choice, including providing value to customers when they add value to the system.
- Argues that bottom-up optimisation via decentralised energy resources is more cost effective and resilient than traditional top-down centralised operation.
- Uses administered/regulated programs (i.e. to support renewable energy, energy efficiency measures etc.) to develop the necessary building blocks for efficient market activity, and then envisages the administered programs decline in importance relative to markets as, for example, RE becomes competitive or as companies delivering EE services become more mature etc.

NY REV building on a decade of supportive policies

The REV has not come out of nowhere. NYS has had a decade of sustainable energy policies. It now has a comprehensive NYS Energy Plan (SEP) (NYS 2015) which (so far continues to) fit with the Obama US Clean Power Plan, NYS has a Greenhouse Gas (GHG) target for a 40 per cent reduction from 1990 levels by 2030 (and by 80 per cent by 2050). NYS also has multiple administered/regulated programs (NYS 2015). For example, a Clean Energy Standard (Cuomo 2016a) which has a target to generate 50 per cent of electricity from clean and renewables by 2030 (NYISO 2016); an energy efficiency program via a Clean Energy Fund (CEF, CEF Order (NYS)) and an Affordability policy (Cuomo 2016b).

However, NYS is behind CA both in terms of sustainable energy capacity implementation, and in terms of length of time that serious sustainable energy policies have been in place. For example, in 2018, solar deployment in NYS had increased by orders of magnitude from 2010 to 1.3 GW, well under the GW 22 of solar that CA had (in 2018) (SEIA 2018). In 2016, for the first time, New York obtained more than 1 million MWh of electricity from solar generation, and 84 per cent of that power came from distributed sources such as rooftop solar panels, New York obtained 24 per cent of its electricity from renewable sources in 2016 (US EIA 2018a).

An upside of the newness of the NY REV is that it is less constrained than CA by historical developments, and the momentum /inertia developed by 'ways of doing things'. Transforming to a new energy system is still in the early stages in NYS, and as a result, NYS has been able to be innovative in deciding where it wants to get to; what the underlying philosophy should be; and what is the best process for delivering the Vision. The NY state government, while setting the overall goals, has been less controlling of the PSC than is the case in CA, where the state effectively sets the decisions that the CPUC (and other agencies) follow through with.

An upfront and clear vision for change

The NY PSC's Vision, as described above, has been upfront in its questioning of two central assumptions of the traditional utility paradigm: (1) that there is little or no role for customers to play in addressing system needs; and (2) that the centralised generation and bulk transmission model is invariably more cost effective than decentralised, due to economies of scale. Thus, the NY REV rhetoric argues that the current business-as-usual regulation cannot deal with the current challenges that energy systems face, nor can it capture the available opportunities. This is a very 'upfront' and challenging Vision.

The intention of the Vision was to find a 'new' energy system paradigm (and its institutions and actors) which suits the current challenges that energy systems face (and NYS in particular), and which can capture its opportunities for the benefit of NYS customers and the NYS economy. It was from the start, therefore, open to a new means of regulation; a 'new' balance between regulation and markets; a new role for actors within the energy system, including the regulator itself, as well as the utilities and customers.

The NY REV envisages an evolutionary rather than revolutionary transition and the 2015 Market Design and Platform Technology Report (MDPT 2015) set out 'an end' of where it expects the NY REV to get to after about 10 years (or the mid-2020s). This is a smart, primarily decentralised, market/platform-based transactive energy system with a new basis of regulation; new institutions; new roles of central actors; and new ways of making money.

Whether, the NY PSC would be so upfront, with hindsight after three years, about the need for change is an interesting question. On the one hand, it signposted where it wanted to go and achieved support from a wide range of stakeholders, including NYS citizens. It also encouraged new companies to come in to the market – for example, the Brooklyn micro-grid (a Peer to Peer (P2P) blockchain using platform[43]) On the other hand, it also immediately signposted

[43] http://brooklynmicrogrid.com/.

to utilities (and utilities in the wider world) that change was coming – and therefore set them up to, at best, be wary.

The transformational institution of the 2014 NY REV, expanded on by the 2015 MDPT, was that of the role of distribution system providers (DSPs). The six main distribution utilities within NYS are expected to transform to DSPs – essentially distribution energy and system service market facilitators and coordinators – who would also be responsible for the public service obligation (PSO). The MDPT report set out the argument behind why utilities need to understand their area distribution energy resources (DER) in detail; the requirement for them to produce a distribution system implementation plan (DSIP); and an explanation of what a DSIP should consist of, what methodology to use when writing one, and how long it would take (about 2 years). The writing of the DSIPs is the process which essentially forces the utility, and other stakeholders (i.e. providers of resource, customers, new entrants, the utility itself), to understand at the most detailed level what the implications are for all stakeholders of being in a regulatory environment which is transforming the distribution utility environment into a DSP environment (NYDPS 2016). Having understood the DER resource in an area, the DSIP was to provide confidence to the wider community that DSPs would, over the 10 year period, develop into:

- DSP platform/markets to support 3rd party investment in DERs;
- benefit utility customers by reducing overall electric system costs and provide them with new services;
- 'animate' the distribution level markets through various mechanisms; and
- provide efficient, linked role between the DSPs and NYISO.

The 2015 Market Design and Platform Technology Report also kicked off various working groups to work out various technical and economic issues of the NY REV, which included in 2018 a Comprehensive Energy Efficiency Initiative (NYREV 2018).

On 19 May 2016, a White Paper on Ratemaking (two years after the NY REV kick-off) was published which sets out the different ways that utilities will be expected to make money in the DSP future, and the timelines for doing so. In general, the Ratemaking Order (SNYPSC 2016a) made less difference to utility revenue over the next 3 years than might have been expected from the rhetoric of the 2014 Vision.[44] However, it did show that the way of regulating utilities and the revenue base of the utility is expected to fundamentally change over

[44] The EAMs should not add more than 2 per cent of their delivered revenues to the distribution companies in the first phase (i.e. over the next 3 years). Given that current performance-based regulation is related to slightly less than 6 per cent of total delivered revenues in NY, this takes PBR up to about 8 per cent, which is a small change.

time; and provided a means for linking the utility revenue base and wider NY state energy program (Mitchell 2016).[45,]

A new world for utilities

The Ratemaking Order represents a new balance of regulation and markets for the distribution utilities where the respective role of regulation and markets is different, and the ways of obtaining revenue also changes. The three main ways of making money will change over the 10 or 15 years from being mainly from traditional cost of service to partly:

- A continuation of the usual: money will continue to be made through traditional cost of service
- a new bringing together of regulated performance incentives mechanisms to fulfil Government policies through Earning Adjustment Mechanisms (EAMs) – by helping to enable public policies and goals are delivered (for example, the renewable energy or energy efficiency programs); and
- a new performance-based regulation where revenues are accrued through platform service revenues (PSRs), which are linked to the developing transactive energy markets.

Markets are central to this new REV Vision. Innovative, local markets/platforms and new ways to operate the energy system are already being stimulated in NYS. However, whilst markets are central, so is regulation. The NY REV took a philosophical choice that distribution utilities would continue with this public service requirement. The NY REV therefore is an experiment in:

- Facing the new utility challenges, including the increasingly serious issues of how to pay for networks given changing technologies and means of provision,
- traditional concerns of ensuring vulnerable customers remain able to access affordable energy; and
- trying to encourage innovation and new behaviours via markets.

Together, as shown in Figure 18, this is transforming the revenues and activities of the distribution companies. This is discussed further below.

Valuing distributed energy resources

In mid-November 2016, the methodology for valuation of DER in NYS (VDER) (SNYPSC 2016b) was published and explained (1) that the current support for DER, mostly through net energy metering (NEM), would continue for a period

[45] See Mitchell at: http://projects.exeter.ac.uk/igov/wp-content/uploads/2016/04/Distribution-Service-Providers-Update-November-2016.pdf.

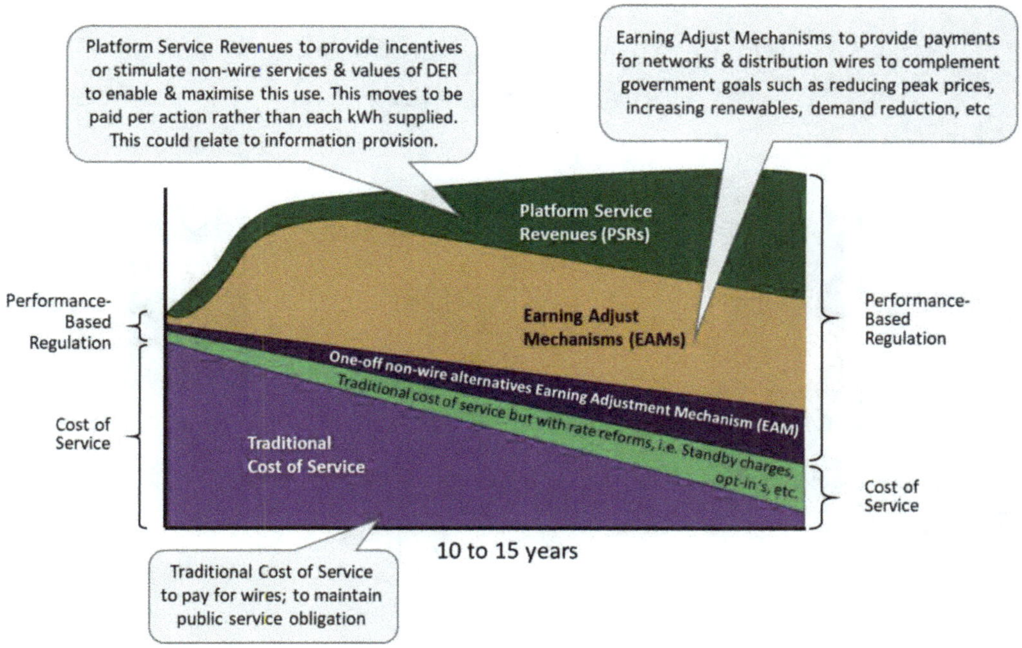

Figure 18: Sources of utility revenue (rate of return on equity) over time within NY REV.

Source: Own contribution.

of time; but that (2) support for NEM would gradually move to a new valuation methodology of DER, which would take account of various aspects. The VDER 4 basic components and known as the value stack (Walton 2017), are:

- the base locational marginal price of energy (LMP);
- the capacity value;
- the credit for the environmental benefits of carbon-free, distributed energy resources;
- a market transition credit.

The VDER agreement to transition away from retail rate net metering for residential customers is largely unprecedented in the United States – it being a major source of contention in most States which have increasing proportions of solar energy. Dubbed the 'Solar Progress Partnership (Shallenberger 2016)', the NYS deal brought together members from both sides (Yahoo Finance 2016) of the distribution grid, including Consolidated Edison, SolarCity, SunEdison, Central Hudson Gas and Electric, New York State Electric and Gas, National Grid, Rochester Gas and Electric, Orange and Rockland Utilities, and SunPower Corp.

The VDER market transition credit in the VDER was a placeholder for the value of the distribution level benefits which DER provides. Having established the methodology in the November 2016 VDER, and taken account of

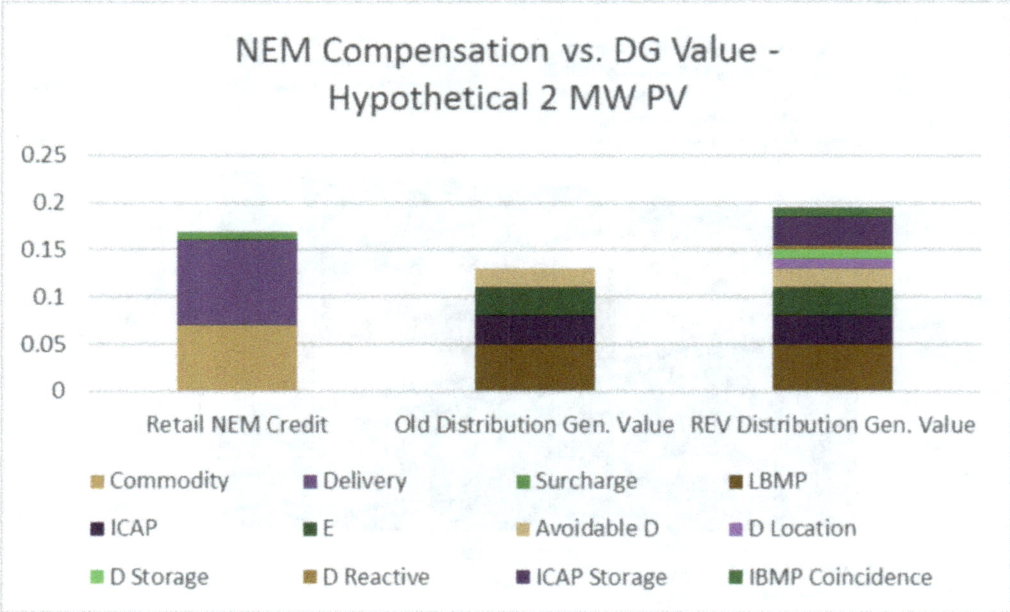

Figure 19: The Value Stack of Distributed Energy Resources in NYS.

Source: SNYDPS (2016a).

Notes:

1. *The 'retail NEW Credit' column represents compensation NEW provides per kWh*
2. *The 'Old Distributed Gen. Value' column represents the potential value that maybe provided under NEW prices signals where the kWh and kW benefits are calculated and then expressed on a per kWh basis*
3. *The 'REV Distributed Gen. Value' represents the potential locational kW and kWh value that could be created if NEM prices signs are replaced with most efficient prices signals.*

the various MDPT projects, practical steps valuing DER moved forward in March 2017, with three dockets from the PSC (Baldwin Auck 2017): (1) an introduction to the Value of Distributed Energy Resources,[46] (2) Distributed System Implementation Plans,[47] and (3) the Interconnection Earnings Adjustment Mechanisms.[48] All these Orders are more detailed workings of previous more generally stated agreements, and will be central to how different entities, including distribution utilities, can make money (Stein & Ucar 2018).

---

[46] http://documents.dps.ny.gov/public/MatterManagement/CaseMaster.aspx?MatterCaseNo=15-E-0751.

[47] http://documents.dps.ny.gov/public/MatterManagement/CaseMaster.aspx?MatterSeq=44991.

[48] http://documents.dps.ny.gov/public/MatterManagement/CaseMaster.aspx?MatterSeq=44991.

The NY REV distribution system provider – the new heart of the energy system

The NY REV argued for a new value proposition, or 'heart' of the newly envisioned NYS energy system – a distribution system provider (DSP). The idea of a DSP is transformational in the sense that it is a new function:

- able to coordinate system operation and balance markets within their service areas;
- whilst at the same time working towards meeting state policy goals
- regulated via PBR so that incentivises new behaviours of the DSP itself, its customers and its service providers, turning the current 'passive' distribution utilities into 'active' market facilitators and system coordinators.

As Figure 19 above shows, revenue comes from three main sources (two of them new): traditional cost of service, EAMs (which help meet state goals); and PSRs (from transactions – as many as possible market based and involving new entrants and new services).

All of this depends on a detailed understanding of the value of distributed energy resources in the distribution area. Only when those values are understood can the DSP coordinate and balance the area as cost effectively as possible. This is why the DER plan was instituted; why the value stack was created; and why a placeholder for DER value was established.

The DSP is the institution which will dynamically be able to value DER. As the system operation situation changes in any one place at any one time – for example, because of EV take up; increased energy efficiency; more DSR; more DER etc. – then it, as the coordinator can promote new transactions which best suit customer/user wishes. This is unlike the CA DRP which is static and coordinated by the monopoly utility.

The DSP in NYS is envisaged as a combined energy and system services market facilitator; the combined wires and energy local system operator and balancer, and provider of last resort. It is the coordinating and balancing platform for an area – but one which coordinates other third-party providers of DER (which may also have their own platforms) which would sell those 3rd party services (of all types) to customers (of all types) via the new 3rd party markets/platforms to create value for both customers and the system.

This, in theory, allows independent DER to bypass the wholesale market and the transmission operator, thereby creating a new value proposition for decentralised energy and the distribution utilities – all the while revealing a new economics of energy provision. This is opening up potential new market possibilities which the DSP, in theory, could, and according to the NY REV philosophy, should, facilitate; and which they are not in control of.

Thus, on the whole, the DSPs in NYS are expected to be facilitators and coordinators rather than 'doers' themselves. If they want new services in their areas, then they are expected to provide incentives, contracts or stimulate tariffs (and necessary data and analysis) to enable a third party to provide those new services.

But, importantly, if customers/users want to do something, then that would either occur under a 'normal' market situation or they can/should approach the DSP to develop a market or a new means (non-wire option) of providing the service. The idea is that the DSP would be incentivised sufficiently for each transaction it encourages to make the economic choice to do that rather than to continue with conventional wires options. This is different from CA – where CCAs are the nearest way for users to do something that they want – as opposed to being controlled by the utilities – in the system.

There are various on-going discussions within the NY REV evolution about the extent to which the traditional utilities are able to maintain their incumbent competitive advantage over the new entrants and new services which the Vision says it wants to encourage. It is clear that the distribution utilities are going to undergo significant threat to their conventional means of operation. It is too soon to be able to say that the way that the NY REV has been undertaken is the 'right' way for energy system transformation. It can be said, however , that it is illuminating issues which have to be addressed if energy systems are to reduce GHG and become resilient; and it seems right in this very technologically fluid time that system operation and economic regulation is designed to be dynamic and flexible.

2.8.4 Comparison of California and New York energy policies

Similarities

Both CA and NYS are major US states in economic and population terms. They both have very high total energy demand by state ranking (2nd and 4th respectively) and they both have low per capita consumption (49, 3rd lowest) and 51 (the lowest) (US EIA 2018b). They are both trying to ensure vibrant economies whilst at the same time meeting their progressive energy policies. CA has had a longer-term, supportive policy for sustainable energy than NY, but both have a reasonably environmentally literate public that has been slowly built over time. Moreover, NYS is very proud of its energy history – being the home of the first Edison power plant. They both take an enabling environment approach to their policies where legislated policies, such as support for renewable energy, energy efficiency etc., intertwine with pro-environmental transport, security and affordability issues. They are both moving towards a greater importance of the distribution level within their energy systems, in part due to the decentralising opportunities of energy technologies; they both support integration between electricity, buildings and transport; and they both support a move to a flexible, smart future. Both have political buy-in for their governance changes.

In their different ways, both CA and NYS have had events which have shaped the development of their energy policies. In CA, the failure of the electricity privatisation has led to wariness of, and difficulty in moving towards, markets – and therefore a reluctance of moving away from the traditional regulatory model. The CA State Legislature continues, effectively, to mandate change and the various CA agencies then implement it. In NYS, Hurricane Sandy has led to openness for change, and a questioning of the 'old' way of doing things. CA was an early mover with respect to energy system transformation and has therefore got various institutions and ways of doing things in place. Hurricane Sandy effectively allowed the past to be reinvented in NYS.

Differences

CA and NYS do, however, have very different approaches to how they communicate and execute those policies. NYS was upfront in the PSC 2014 Vision arguing that traditional energy regulation is no longer fit for purpose, and that new roles for utilities, customers, and the Regulator are required. Moreover, NYS has been clear that it sees a rebalancing between Regulation and Markets as necessary – with both regulation and markets as having important roles: Regulation ensures direction and maintains a public service obligation whilst more markets enable innovation, cost effectiveness, and customer choice.

CA has not been as strident in its public pronouncements. CA still effectively delegates executive responsibility for its sustainable policies to its IOUs, which then generally procure renewables etc. for sale to their customers. Regulation remains the more important aspect of the alternative tools of regulation or markets. The role of the different actors in energy system transformation have not been so openly questioned in CA. Whilst CCAs are making some inroads in CA, most domestic customers remain served by the incumbent IOUs, and even CCAs are reliant on the IOUs for billing and so forth.

Whilst NYS utilities remain at the centre of the energy system transformation, the basis of their future revenues is set to alter fundamentally, and part of that revenue is related to transforming institutions (the DSP) and enabling new players and new transactions. So far, whilst it is clear to all concerned that energy systems are having to grapple with all sorts of change, including the roles and business models of those involved, the CA Regulator has not added details of how the basis of the utilities revenue will change over the next decade or so. In this sense, the environment for the CA IOUs is far less threatening than in NYS.

NYS has also been clear that utilities can earn more if they change – so there is a strong carrot and stick element within the NY REV, which in theory should be attractive to utilities. At the same time, in theory, new entrants and new ways of doing things should be more attractive in the NY REV world of DSPs and markets and, certainly, it seems that the ideas are squarely positioning NYS to complement D4 and transactive energy. However, CA continues to be the major market for DER in the USA, and whilst it may be a utility which procures, for example, DSR or solar from new entrants because of regulation rather than from via a market in NYS, CA is still providing greater amounts of value to DER than NYS. From a practical perspective therefore, whilst new entrants may like the NYS philosophy, it is still CA which is coming through with value which can be captured.

Nevertheless, while utilities may prosper in the new world, systems are also moving quickly and forcing issues upon all system stakeholders. CA has had negative prices in electricity markets for the first time – a sign that the system has moved to a new existential phase (US EIA 2017).

The NY REV has engendered a much more an in-depth, transparent discussion than CA about what the role of the energy system should be in the 21st century energy system, and what would be appropriate regulation. NYS has tried to open up the debate about what energy policy is good for the NYS economy, as well as what customers want from the energy system and the ways in which it could connect customers, including domestic customers, to their energy use. This has shifted the debate from the interests of companies to one of public interest and resilience. Moreover, at its very roots, the NY REV is de facto arguing that an energy policy built around public interest is likely to be more successful in meetings its public policy goals than the old 'private' interest model – this is both radical but also very different from CA.

NYS is also arguing that customers will still need protection in this new world whilst placing them and their customer propositions at the centre of the energy system service. This is exploring new ground. Whilst it is the same century old mandate, it is a new compact with customers. The NY PSC has opted, at this time, to say that the utilities have to continue with this public service obligation. New entrants, and new ways of providing services, currently do not offer an alternative to the regulated route of public service provision. The rapid changes of the energy system are exciting but the NY PSC has said, much as they support and want to encourage new markets and new roles, customers, particularly vulnerable customers, have to be looked after. In this way, the NY PSC is marrying a traditional regulatory role with the opening up of markets and new forms of performance based regulation via EAMS and PSRs. It is this triad approach, which together is unique and interesting.

NYS has been able to capture the position of regulatory innovator because it is the first state to put forward – and take steps to execute – such innovative ideas, centring on the Distribution System Providers for DER. CA has implemented much of the same actions around distribution development – for example, DRPs, Action plans, and ways to value DERs – but it is implementing them a very traditional way.

Nevertheless, both CA agencies and the NY PSC are cautious when it comes to dealing with the distribution utilities in their respective states. There are concerns on the part of the distribution companies about the impacts of the transformations on their businesses and neither CA nor NYS are pushing the utilities too hard, as yet, for change. Moreover, whilst the NY PSC continues to see the utilities as providing the public service of final resort, the utility future still seems reasonably assured.

The transformational nature of ideas

NYS appears to be trying to create a process where either regulation or markets can be chosen as the tool to reach a desired outcome, depending on which is best for each desired outcome. Their arguments in support of local balancing markets, platforms, DSPs as coordinators, performance incentives and so on are new institutions and new regulations coming together in new ways. The idea is that this will lead to a more cost effective regulation which provides the services and propositions that customers actually want.

The unknowable issue, and in a sense the most important question, for States or Countries which are thinking about transforming their energy systems, is whether the more regulated approach of CA is preferable, cheaper, quicker, easier in moving those States or countries towards GHG reduction than the NYS model. Whilst NYS is arguing that innovation and a customer focus is necessary in order to achieve a cost effective energy system transformation – and that markets, the decentralised value proposition of DSPs, and performance based regulation is integral to that – we cannot as yet know if that is the case. NYS is placing customers at the centre of the energy system, and wants the services that they want to be provided – and incentivised via performance based regulation. In other words, NYS is attempting to move beyond a narrow cost effective service for individual anonymised customers towards enabling individual customer choice – of any type – to buy and sell to and from whomever they wish to.

This is a fundamentally different approach. From a theoretical point of view, this chapter prefers the NY REV model to avowedly free up innovation to develop in ways 'it' (the innovation) wishes to and which 'we' (the regulator/industry) cannot know about now. The chapter also in theory agrees that a DSP like value proposition is best able to work with, and coordinate, with D4 and the developing transactive energy platforms, and other changes. Moreover, the NYS DSIP and its valuation methodology, being dynamic, seems to be preferable.

This chapter is impressed that NYS has attempted the NY REV. It recognises that NYS has undertaken a major program in trying to deliver new regulatory approaches across a number of fronts. This may prove to be too big a step, in which case the CA approach may turn out to be preferable in terms of on the ground, GHG reducing, practice change. On the other hand, the NY REV may all come together in a few years to meet the challenges of D4 and to take up its opportunities in a way that CA may find itself unable, or constrained, to do because of its lack of customer choice.

It also seems that some of the lessons learnt from Europe could also be beneficially incorporated into the NY REV – and that includes separating out distribution utilities from 'supply': in other words, turning DSPs into energy and wires companies but removing their supply base.

Data, and its availability as a public good versus a source of revenue is also being hotly debated in Europe, as it already is in NYS. This chapter takes the view that data should be freely available and seen as a public good. Another similar debate in Europe and NYS, is whether DSPs should own or not own DER. This chapter takes the view that it should not.

It is true that the changes on the ground in NYS itself have not occurred as fast as some would hope. However, if the 2014 Vision and principles are separated from the on-the-ground change in NYS, the NY REV is offering a new answer to the challenges of the 21st Century. In this sense, it is inspirational. If the goal is energy system change – then it is a case of 'watch this space' to see just how fast NYS is able to alter, and what the problems and difficulties have been so that lessons can be learned.

This is not to in anyway undervalue CAs history to date – which is clearly the most successful in the USA, and one of the most successful in the world in terms of RE deployment and decarbonisation of mobility.

Finally, in conclusion, this chapter argues that a combination of steady public policy targets and support (as has occurred in CA for 40 or so years, and for a shorter period in NYS) combined with the new institutions, centrality of customers; balance between regulation and markets; and new regulatory incentive mechanisms in NYS are, at this time, the 'best practice' lessons coming out of both CA and NYS governance for DER.

Author's note

A small amount of the above chapter was previously published by the author on the IGov New Thinking for Energy blog, found at http://projects.exeter.ac.uk/igov/comparing-nys-with-ca-blog-4-a-comparison-of-the-fundamental-regulatory-principles/

2.8.5 References

Baldwin Auck, S. (2017), *New York regulators issue three milestone REV orders, but more work remains. GTM, 15 March 2017.* Retrieved from https://www.greentechmedia.com/articles/read/New-York-Utility-Reformation-Is-Hard?utm_source=twitter&utm_medium=social&utm_campaign=gtmsocial#gs.TmjIl_I

Brockway, A.M. (2017), *Distributed generation planning: a case study comparison of California and New York proceedings. Center for Sustainable Energy.* Retrieved from https://energycenter.org/sites/default/files/docs/nav/policy/research-and-reports/Distributed_Generation_Planning.pdf

CAISA. (2018), *Distributed energy resource provider, Californian independent system operator.* Retrieved July 2, 2018, from https://www.caiso.com/participate/Pages/DistributedEnergyResourceProvider/Default.aspx

CAISA. (2014a), *Advancing and maximizing the value of energy storage technology – a Californian roadmap, Californian independent system operator, December 2014.* Retrieved from https://www.caiso.com/Documents/Advancing-MaximizingValueofEnergyStorageTechnology_CaliforniaRoadmap.pdf

CAISA. (2014b), *Relevant CPUC, energy commission, and ISO proceedings and initiatives, - California Energy storage roadmap companion document December 2014.* Retrieved from http://www.caiso.com/Documents/CompanionDocument_CaliforniaEnergyStorageRoadmap.pdf

California State. (2006), *SB-32 California global warming solutions act of 2006: emissions limit.* Retrieved from https://leginfo.legislature.ca.gov/faces/billNavClient.xhtml?bill_id=201520160SB32

California State. (2018) *Assembly Bill 32 overview.* Retrieved July 2, 2018, from https://www.arb.ca.gov/cc/ab32/ab32.htm

California State. (2016), *California's distributed energy resources action plan: aligning vision and action, November 2016.* Retrieved from http://www.cpuc.ca.gov/uploadedFiles/CPUC_Public_Website/Content/About_Us/Organization/Commissioners/Michael_J._Picker/2016 percent20DER percent20Action percent20Plan percent20FINAL.pdf

CEC. (2008), *Energy action plan, 2008 update, Californian Energy Commission.* Retrieved from http://www.energy.ca.gov/2008publications/CEC-100-2008-001/CEC-100-2008-001.PDF

CEC. (2018), *Tracking progress. Californian energy commission.* Retrieved from http://www.energy.ca.gov/renewables/tracking_progress/documents/renewable.pdf

CEC. (2018), *Final 2017, integrated energy policy report. Californian Energy Commission, April 2018.* Retrieved from see http://www.energy.ca.gov/2017_energypolicy/

CPUC. (2012), *Community choice aggregation allows communities to offer procurement service to electric customers within their boundaries - Code of Conduct D12-12-036. California Public Utilities Commission 20 December 2012.* Retrieved from http://www.cpuc.ca.gov/general.aspx?id=2567

CPUC. (2013), *Distribution resources plan (R.14-08-013), California Public Utilities Commission.* Retrieved from http://www.cpuc.ca.gov/General.aspx?id=5071

CPUC. (2014), *Order instituting rulemaking regarding policies, procedures and rules for development of distribution resources plans pursuant to public utilities code section 769. California Public Utilities Commission.* Retrieved from http://docs.cpuc.ca.gov/PublishedDocs/Published/G000/M103/K223/103223470.pdf

CPUC. (2016), *California's distributed energy resources action plan: aligning vision and action, 10 November 2016.* Retrieved from http://www.cpuc.ca.gov/uploadedFiles/CPUC_Public_Website/Content/About_Us/Organization/Commissioners/Michael_J._Picker/2016 percent20DER percent20Action percent20Plan percent20FINAL.pdf

Cuomo. (2016a), Governor Cuomo announces establishment of clean energy standard that mandates 50 percent renewables by 2030. New York State, 1 August 2016. Retrieved from https://www.governor.ny.gov/news/governor-cuomo-announces-establishment-clean-energy-standard-mandates-50-percent-renewables

Cuomo. (2016b), *Governor Cuomo announces new energy affordability policy to deliver relief to nearly 2 million low-income New Yorkers. New York State, 19 May 2016.* Retrieved from https://www.governor.ny.gov/news/governor-cuomo-announces-new-energy-affordability-policy-deliver-relief-nearly-2-million-low

DRPWG. (2013), *California's distribution resources plan. Distribution Resources Plan Working Group.* Retrieved from https://drpwg.org/sample-page/drp/

Lacy, S. (2014), *Watch the energy gang's live podcast on utility 2.0. GTM, 21 September 2014.* Retrieved from https://www.greentechmedia.com/articles/read/join-the-energy-gang-for-a-live-podcast-in-new-york-on-utility-2-0#gs.vJtV3Bg

MDPT. (2015), *Report of the market design and platform technology working group. NY Smart Grid, 17 August 2015.* Retrieved from http://nyssmartgrid.com/wp-content/uploads/MDPT-Report_150817_Final.pdf

Mitchell, C. (2016), *US regulatory reform series – what, and how, the New York utilities are expected to transform to over the next decade – the New York REV's Ratemaking May 2016 Order. Energy Policy Group, University of Exeter, 13 June 2016.* Retrieved from http://projects.exeter.ac.uk/igov/us-regulatory-reform-ny-utility-transformation/

Navigant Research. (2017), *Navigating the energy transformation.* Retrieved from https://www.navigantresearch.com/research/defining-the-digital-future-of-utilities

NYDPS. (2016), *Staff report and recommendations in the value of distributed energy resources proceeding 15 E-0751 NY Department of Public Service, 27 October 2016.* Retrieved from https://s3.amazonaws.com/dive_static/editorial/Staff+Report+and+Recommendations+10-27+(1).pdf

NYISO. (2016), *NYISO issues power trends 2016, report surveys the changing energy landscape, NY independent system operator, 5 July 2016.* Retrieved from http://www.nyiso.com/public/webdocs/media_room/press_releases/2016/NYISO percent20Issues percent20Power percent20Trends percent 202016_7-05-2016.pdf

NYREV. (2014), *Reforming the energy vision NYS Department of Public Service staff report and proposal, 24 April 2014, NYS Department of Public Service.* Retrieved from http://www3.dps.ny.gov/W/PSCWeb.nsf/PFPage/C12C0A18F55877E785257E6F005D533E?OpenDocument

NYREV. (2018), *DPS- reforming the energy vision.* Retrieved July 2, 2018, from http://www3.dps.ny.gov/W/PSCWeb.nsf/All/C12C0A18F55877E785257E6F005D533E?OpenDocument

NYS. (2018) Clean energy fund, New York State. Retrieved July 2, 2018, from https://www.nyserda.ny.gov/About/Clean-Energy-Fund

NYS. (2015), *The energy to lead, 2015 New York state energy plan. New York State*. Retrieved from https://energyplan.ny.gov/Plans/2015

RENSWICK Institute. (2015, June), More than smart, a framework to make the distribution grid more open, efficient and resilient. *Gridworks*. Retrieved from http://gridworks.org/wp-content/uploads/2015/06/More-Than-Smart-Report-by-GTLG-and-Caltech-08.11.14.pdf

SEIA. (2018), *Solar state by state, Solar Energy Industry Association, 2018*. Retrieved from https://www.seia.org/states-map

Shallenberger, K. (2016, May 26), Strange bedfellows: how solar and utilities struck a net metering compromise in New York. *Utility Dive*. Retrieved from https://www.utilitydive.com/news/strange-bedfellows-how-solar-and-utilities-struck-a-net-metering-compromis/419367/

SNYPSC. (2016a), *Order adopting a ratemaking and utility revenue model policy framework, State of New York Public Service Commission, May 2016*. Retrieved from http://documents.dps.ny.gov/public/Common/ViewDoc.aspx?DocRefId= percent7BD6EC8F0B-6141-4A82-A857-B79CF0A71BF0 percent7D

SNYPSC. (2016b), *Staff report and recommendations in the value of distributed energy resources proceeding, State of New York Public Service Commission, October 2016*. Retrieved from https://s3.amazonaws.com/dive_static/editorial/Staff+Report+and+Recommendations+10-27+(1).pdf

Stein, E. &Ucar, F. (2018), *Driving environmental outcomes through utility reform: lessons from the NY REV. Environmental Defense Fund, January 2018*. Retrieved from https://www.edf.org/sites/default/files/documents/driving-environmental-outcomes.pdf

Sweeney, J.L. (2002), California electricity crisis. Stanford, CA: Hoover Institution Press.

Trabish, H.K. (2017, May 9), Choice in la l: LA county community aggregation has California utilities on full alert. *Utility Dive*. Retrieved from https://www.utilitydive.com/news/choice-in-la-la-land-la-county-community-aggregation-has-california-utilit/442131/

US EIA. (2017), *Rising solar generation in California coincides with negative wholesale electricity prices. US Energy Information Administration, 7 April 2017*. Retrieved from https://www.eia.gov/todayinenergy/detail.php?id=30692#

US EIA. (2018a), *New York state profile and energy estimates, US Energy Information Administration 2018*. Retrieved July 2, 2018, from https://www.eia.gov/state/?sid=NY#tabs-4

US EIA. (2018b), *US energy mapping system, US Energy Information Administration 2018*. Retrieved July 2, 2018, from https://www.eia.gov/state/maps.php?src=home-f3

Walton, R. (2017, March 15), New York REV orders promise growth for diverse set of distributed resources. *Utility Dive*. Retrieved from https://www.utilitydive.com/news/new-york-rev-orders-promise-growth-for-diverse-set-of-distributed-resources/438044/

Wesoff, E. (2017, March 15), California's energy future: the revolution (might) be distributed. *GTM*. Retrieved from https://www.greentechmedia.com/articles/read/Californias-Energy-Future-The-Revolution-Might-Be-Distributed?utm_source=twitter&utm_medium=social&utm_campaign=gtmsocial#gs.jWNvr5M

Yahoo Finance. (2016), *NY energy utilities and solar providers file joint proposal to encourage more renewables, 19 April 2016*. Retrieved from https://finance.yahoo.com/news/ny-energy-utilities-solar-providers-195725145.html?guccounter=1

2.9 Conclusions and reflections from the country reports

The last decade has witnessed the beginning of what is likely to be a fundamental, irreversible transformation of the power and wider energy sectors, on a scale similar to that seen in information and communication technologies with the rise of the desktop PC and the smartphone. This is primarily because of decarbonisation and digitalisation. Digitalisation is part of the General Purpose Technology (GPT) family which has now effectively come to energy. When combining this with decarbonisation – a general societal transformation of the same sort as a GPT but in policy terms – a fundamental technological and societal shift is unleashed towards decentralisation.

These forces also suit differing country needs. For example, in India decentralisation fulfils goals of access, reduced pollution, domestic jobs and decarbonisation. In China, it supports regional development and technological growth as well as decarbonisation. In the United States, it fosters state independence from federal policies along with multiple state-based priorities, whether energy independence, security, GHG reduction, air pollution reduction, and so on. In Europe, it helps meeting climate change objectives and industrial strategies. In Australia, and an increasing number of countries and regions, the rapid growth of decentralisation is linked to the competitiveness of renewable energies under favourable climatic conditions – households can buy cheaper energy via onsite technologies than from suppliers.

With the increased deployment of renewables, primarily solar and wind, renewable generation has become the dominating investment opportunity globally, and changes the way in which the sector operates.

Enabling a system that can efficiently integrate these new technologies implies a fundamental change in energy governance. Most regulatory frameworks have been designed to secure reliable operations of the centralised power system, but they may not have changed sufficiently to reflect the imperative to meet internationally agreed decarbonisation objectives. Change may be triggered by climatic events, such as Hurricane Sandy in New York, or storms and heat waves in Australia. However, all country reports suggest that governance is a decisive factor in the successful process of the transformation, as **governance can act as an accelerator or decelerator of the transformation.**

The difficult task for regulators and policy makers is facilitating a rapid but smooth transformation from the 'old' energy system to the 'new' in a dynamic technological and economic environment.

How to cite this book chapter:

Burger, C., Froggatt, A., Mitchell, C. and Weinmann, J. 2020. *Decentralised Energy — a Global Game Changer.* Pp. 157–175. London: Ubiquity Press. DOI: https://doi.org/10.5334/bcf.j. License: CC-BY 4.0

2.9.1 The eight requirements of rapid transformative governance

At the beginning of this chapter, we provided an overview of the key reasons for why these countries were reviewed to assess their potential and actual pathways to decentralisation. Derived from the country reports, eight areas that require political action are identified, and these are discussed below:

1. Transparency and ligitimate policy making and institutions.
2. Availability and transparency of data.
3. Customer focus, enabling customer choice.
4. Markets to encourage flexibility in supply and demand.
5. Local system coordinators and a coexistence of the central grid and decentralised micro-grids.
6. Reforming regulation – Including performance-based elements.
7. Reassessing investments in the long-distance transmission grid.
8. An integrated approached to sector regulation.

Transparency and legitimate policymaking and institutions

A transition towards a low-carbon decentralised energy system can take many technological pathways, and the distributional impacts of those choices can be both positive and negative on societal stakeholders. Governance mechanisms of the energy system transformation are most likely to be supported by the general public if they are 'legitimate' – and one important characteristic of legitimate policy making is transparency in how decisions are made. If the policy-making process is flawed by, for example, governmental institutions undertaking myopic regulatory decisions, pursuing short-term political interests, decision makers underestimating the complexity of the system, or corporations successfully lobbying for their particular interests, legitimacy and authority may be jeopardised, and a regulatory regime may emerge that excludes alternatives which might suit social interests and preferences better.

Some countries have established a consensus culture, entailing several dimensions: the actual political process of voting in governments, proportional representation versus First Past the Post (such as in the UK) the degree to which decisions are devolved, for example, federal versus state in the United States and Germany; and the extent to which local areas can in some way make their wishes known;[49] the degree to which a society is knowledgeable about issues and so can meaningfully decide about them; the degree to which efforts are made by a society to ensure that society members are able to express their wishes, and the desired customer proposition is put in place. Transparent decision-making processes enhance public support for politicians and the overall transformation.

---

[49] For the power sector, this translates into a possibility of buying of local networks, as it happened in Hamburg, Germany.

For example, one of the principles of California's energy policy is to ensure that the state's implementation of these policies is transparent and equitable.

Denmark might be seen as a country that tries to find consensus in multiple ways, whereas other countries, for example Italy or Germany, may have implemented to some extent a consensus culture, or try to do this in a very limited manner, such as China.

In political practice, creating 'consensus' is a conflict-laden road whatever the country. Countries have to establish structures and institutions to be able to deal with conflicts. For example, Danish governance is a set of rules and processes that enable conflicts to end up in solutions. This can be linked to the Danish parliament with its many parties, where the large parties need a smaller coalition partner to get majority in parliament. In effect, proportional representation in a national parliament may be more suitable for constructively handling conflicts than largely two-party regimes. However, a proportional system may lead to the fragmentation of political parties and create political instability or paralysis, as can be seen in Italy's politics in the 1980s.

Government and communities need capacity – institutions, financial resources, human agency – to encourage consensus and understanding of what society wants:

- **Stakeholder involvement:** One way to deliver 'legitimate' decision making is to ensure a process that is designed to be transparent, coherent, and to deliver an acceptable consensus. This requires listening to as many stakeholders as possible and keeping up to date with information about societal preferences, not just economics. For example, in the United Kingdom the Committee on Climate Change (CCC) is statutorily required to provide advice to the UK government on what level of GHG emissions there should be, and by when. However, whilst the government must provide a written reply to the advice, there are no direct requirements on actors within the energy system for themselves to work towards the CCC outputs, despite the importance of their recommendations to the energy system and the importance of the energy system in meeting the recommendations of the CCC.
- **Government involvement:** Energy regulation needs to be recognised and voted on by elected representatives to ensure legitimacy and consumer acceptance, rather than delegating decisions to an independent regulatory body. The distributional impacts of any policy will be different. For example, an energy policy that includes nuclear power as a decarbonisation technology will have very different impacts on different stakeholders than an energy policy without nuclear plants. Making any trade-offs between outcomes that have a significant impact on one particular stakeholder group should not be the responsibility of a regulator or a network company, but should be the direct responsibility of government and consistent with broader public objectives. In addition,

efforts to harmonise national and regional policies may encounter challenges, as it can be observed between the European Union and its Member States. For example, the International Emissions Trading Association (IETA) detects problems within the European Emissions Trading Scheme, whose operation does not consider the success of separate policies encouraging energy efficiency and renewable energy deployment (IETA 2015).

- **Local governance:** A requirement of a smart and flexible energy system is coordination and balancing at the distributed, decentralised level. Energy regulation should encourage the involvement of different actors at different levels. This can be individuals, but also cooperatives or community groups, local authorities or small, decentralised companies through to bigger, more conventional actors, as shown by Denmark, California, New York State, and others. Larger entities can continue to be involved, but the new distributed energy resources require coordination at the local level if the system is to be run efficiently – that will need local governance, which of course needs to fit with the wider energy system governance.

Many traditional electricity generating companies are limited in their ability to transform their operations to keep pace with the changing technology environment. This is not only due to their nature of their existing assets, but also due to lack of familiarity and experience how to handle these new technologies to achieve optimal operational and financial outputs. Consequently, there may be political pressure to support these companies, either by slowing the changes to the regulatory environment, or by introducing specific measures to support their continual existence in the market, or both.

Large utilities may attempt to exert pressure on governments to implement policies and regulatory rules that suit their corporate strategies rather than the public interest, as is highlighted in some of the country reports, including China and Italy (Sections 2.3 and 2.7 of this chapter). Until recently, incumbent generators and large-scale investors in the power sector have had significant influence on development of policies and regulations, for example, seeking assurance and the inclusions into contracts indemnity from policy changes that might have a material impact on their investments. Ensuring the greater engagement by consumers and people in the sector comes with a need for a greater role for them in participatory processes to set regulations, measures, and policy objectives. In many countries this has been recognised by policy makers, and there have been ample opportunities for consumers and people to be consulted on the introduction of new regulations and policy frameworks, including responses to draft regulations or white papers, opportunities to attend 'townhall meetings' and discuss topics such as transmission grid extensions or wind parks in roundtables with various relevant stakeholder groups.

Availability and transparency of data to enable entrants to pursue the route to market of emerging business models

The ability to measure, collect, analyse, and share data has become cheaper and quicker across all aspects of our lives. In the energy sector, this enables individuals to monitor their own consumption in real time, and for companies to immediately act upon it. This has significant implications for the efficiency of grid operations, for generation, distribution and consumption. With access to smart grid data, companies and grid operators can offer better customer services. New private sector entrants, local communities and cooperatives need system and consumer data to be able to figure out whether there is a business case for them to provide a new service. To ensure competition and a level playing field, an entity that acts as a market monitor and data provider makes the data 'freely' available and controls that those with data or market power do not misuse that information.

Some countries or states are more active than others to package data in a more accessible way. As the report in Section 2.8 of this chapter has shown, some jurisdictions move from individualised customer data to system data. Both California and New York states consider assessments of their regulated distributed energy resources. For interested stakeholders, the data may be freely available, based on the argument that this will enable new entrants to understand what the potential of distributed energy resources is in their states.

The introduction of smart meters, whose deployment is driven by domestic or regional policies, allows a first level of information. They cannot, however, work in isolation; rather they can serve as enablers to integrate new actors into the sector, as well as to integrate different segments of the energy value chain into a smarter system. In the EU, the European Commission expects that by 2020, 72 per cent of electricity customers will have a smart meter (European Commission 2017). However, within the European Union there are already and are expected to be significant differences in the deployment rates. For example, by 2020 Denmark and Italy will achieve almost full coverage, while in Germany only 23 per cent market penetration is expected, due to concerns over economic efficiency. China is currently the world leader, with 450 million (from a global total of 700 million). In the United States, more than 60 million households have a smart meter, but adoption rates vary across states. However, privacy concerns, ownership of the data and cybersecurity all pose real threats to the rapid widespread introduction of smart meters (see also Burger, Trbovich & Weinmann 2018). Furthermore, with the ever-increasing opportunities of digitalisation, the fear that newly installed smart meters may become outdated relatively quickly is reducing enthusiasm for wide-scale rollout in parts of Europe. While much of the public is willing to share personal information in other areas of their private lives, in particular communication and social networks, the power industry will have to show the benefits and safeguards for individual consumers.

Furthermore, as private platforms are developed for new resource provisions, for example peer to peer trading, more data is derived which is 'outside of' of the conventional energy system but which will become a larger and more important part of it. Whilst this creates further opportunities for new players in the energy system, it is also likely that there will be a need to understand and resolve fundamental questions regarding data security and the protection of consumer privacy.

Customer focus, enabling customer choice

It seems commonplace nowadays that energy policy is customer-focused. For example, in the European Union the legislation on electricity market reform proposes to put consumers at the centre of the Energy Union; to empower consumers, to provide them with better information on their energy consumption, to make it easier to switch supplier, and to be able to generate and store their own energy (European Commission 2016).

A central determinant of the new energy system is the ability of customers to explore new, revenue-generating opportunities related to their energy use or self-generation. As the prices of supply technologies fall, as governments encourage greater energy efficiency, and as ICT becomes smaller, cheaper and easier to use, more customers are becoming more actively engaged in the energy system. Households, community groups and energy co-operatives, small and medium enterprises (SMEs) and heavy industry turn into producers, who can also play a significant role in balancing the grid (Hoggett 2016).

A combination of factors drives the changing role of consumers in the electricity, and wider, energy system. In many countries with liberalised markets, consumers are able to play an active role in power sector,

- by switching their supplier, choosing a new supplier based on price, fuel mix, ownership structure or combined utility offer, with heating, water or communications;
- by becoming prosumers (producers and consumers) through investment in individual or community level supply options, usually solar or wind;
- by investment in energy infrastructure, such as storage or even grid, as an individual or as part of a community project.

In most countries the rise in the deployment of renewables was accompanied by, and in many countries, such as Australia, Denmark, Germany, Italy, and the United States, driven by small-scale solar and wind deployment financed by private residents – the rise of the prosumers. Increasing customer involvement is seen as a key driver of the energy transformation in many countries (Energy Networks Australia 2017; European Commission 2016). When variable renewables contribute more significantly to overall supply, the success of the transformation will depend on greater engagement of customers in three areas – as investors and operators, as willing participants who pay for the new market and consumers of new products, and as supporters of policies and measures that deliver decarbonisation.

Transactive energy is technically similar to what in Europe goes under a variety of titles including 'Community self-consumption' (France), and 'Tenant self-consumption' (Germany). France, in April 2017, made changes to Article D of its Energy Code to support electricity self-consumption at the grid's edge. Germany has likewise amended the German Renewable Energy Sources Act (EEG 2017) to explicitly include self-consumption of PV electricity by buildings tenants. Both of these anticipated changes foreshadowed the proposed fourth EU Electricity Directive, which substantially enhances measures to proactively support consumer participation in the energy system (Butenko 2018).

In a number of countries, such as China, Denmark, Germany, India, and Italy, households and micro-producers may suffer from disadvantages and discrimination in the electricity system, though. For example, in Germany auctioning and tender systems favour larger suppliers. In countries without liberalised markets, they are not able to choose suppliers, become autonomous, or feed their power into the grid.

China may be indicative of a development trajectory that emerging economies can pursue without customer involvement: While reforms have occurred, with the introduction of new ministries and a move towards independent power production, individual consumers, either through their use of power or their rights, have marginal influence on the power sector. Despite this, the Chinese renewable sector is by far the largest producer and deployer of renewable energy, especially solar and wind. This is driven by top-down targets rather than bottom-up initiatives. This raises questions, can the Chinese system create a long-term, sustainable and engaging power system. How to achieve 'meaningful' consensus has many dimensions – knowledge transfer, education, places for discourse, and decision-making processes that take note of individual preferences.

As the Finkel Review in Australia states:

> 'The retail electricity market must operate effectively and serve consumers' interests. Improved access to data is needed to assist consumers, service providers, system operators and policy makers. Increased use of demand response and changes to the role of networks and how they are incentivised are required to unlock these benefits. Governments also need to take steps to ensure that all consumers, including low income consumers, are able to share in the benefits of new technologies and improved energy efficiency' (Finkel et al. 2017)

There are multiple cases of significant consumer involvement in energy policy:

- Germany offers an example of the success of long-term citizen's empowerment. The beginnings of what is now known as 'Energiewende' date back several decades and have their roots in the oil, nuclear and environmental crisis of the 1970s and 1980s, which resulted in the transformation of energy supply as a bottom-up process. Citizens mobilised significant resistance against the conventional energy policy of those days and activated social engagement for structural changes in the energy policy and supply system. Decades of critical social debates about the existing energy policy led to a counter-proposal to the conventional energy supply, which was adopted by the government in the early 1990s and led to the unprecedented rise of renewable energies.
- The commitment and investment of citizens still remains a key driving force of the German Energiewende: As outlined in the country report in Section 2.5 of this chapter, citizen energy has a market share of 47 per cent of the installed renewable electricity capacity in Germany. Therefore, while public involvement is important, key to longer-term consumer engagement is access to the market. The experience of cooperatives enabled a tested route for citizens to investment and gaining a stake in the sector.
- A similar development has been seen in Denmark where wind power survived on a fragile home market due to a continuation of parliamentary support and subsidies for wind power, which probably would not have prevailed without the policy pressure from the many wind turbine shareholders, including local citizens. As the country report in Section 2.4 of this chapter shows, the conflict regarding the establishment of the new heat and power integration infrastructure leads to the question how the sector can be rooted in a bottom-up and smart energy system to ensure that further integration of renewables is possible without difficulties and unnecessary expense.
- Following the Fukushima accident in Japan in 2011, German citizens expressed increased public and political concern over nuclear power, especially regarding the oldest reactors. This resulted in the introduction of a new direction for energy policy in Germany – the Energiewende (or energy turnaround); to phase out the use of nuclear power, by 2022; to accelerate the deployment of renewables and to increase energy efficiency. This was a dramatic change in the domestic policy, since at the same time the administration had only just introduced legislation to enable the nuclear power plants to continue to operate.
- The nuclear accident in Fukushima also resulted in changes in nuclear power deployment rates in China, the cancellation of new build considerations in Italy, following a referendum; recent announcements of no more nuclear in South Korea; and the cessation of all nuclear power in Japan, with local opposition delaying the restart of many reactors.

- A similar external event led to changes in the electricity sector in New York. Audrey Zibelman, the then chairperson of the NY State Department of Public Service (DPS) described how Hurricane Sandy was a major driver in the NYS decision to articulate a new vision for energy. Hurricane Sandy provided the desire for change of the NYS people, as well as the DPS focus of providing customers with the services they want – which includes security and cost effectiveness.

Customers, together and as individuals, are driving energy policy by their investment decisions, too. This is currently most obvious with respect to residential solar installations – whether in sunny places like Australia or Italy, or non-sunny places like Germany, where the take-up of solar has been far greater than expected and is driving regulatory change. Governments should welcome these murmurating situations, because they have significant benefits from an investment point of view. If customers, individuals or consumer co-operatives, are de facto becoming the investors on the system, there is less need for investors from other sources or – as technology prices fall – state financial support or subsidies. Other potential decentralised murmurations, possibly storage and electric vehicles, may follow. A customer-focused energy policy would be supportive of this murmurations and work with them.

Globally, competitiveness and levelised costs of solar PV and wind are changing, now routinely estimated below operating costs of coal-fired generation. On a retail level, tariffs faced by a vast majority of consumers, for example in India, are substantially higher than the cost of rooftop PV. Non-economic barriers, such as access to credit, are getting addressed by governments and regulatory bodies. Once they are minimised, growth of decentralised renewable energy supply is likely to further accelerate.

Decentralised renewable power can also be part of broader public policy goals by creating local employment, decreasing brain drain from rural areas and urban migration. As the report on China in Section 2.3 states, distributed generation and in particular solar are used as a tool for reducing poverty alleviation. In fully industrialised countries, such as Germany, substantial benefits for local employment have been observed.

On a global scale, significant differences in consumer focus remain, with many people still not having access to any or reliable supply, such as in India; or those that do, many have no choice about their supplier, payment system or tariff, such as in China. In the European context, for example, Italy also experienced a considerable growth in small-scale renewables systems, but policy is still dominated by large supply-side operators.

Designing markets to encourage flexibility in supply and demand

The greater deployment of renewables over the last decade has led to a recognition of the need for a more flexible system in order to accommodate variable producers.

To date four main mechanisms have been suggested to add the necessary flexibility, which are: interconnections, greater flexibility of generation from more predictive plants (often currently promoted through capacity markets or payments), storage, and demand side measures. These mechanisms vary in their suitability according to geographical and economic conditions of energy markets. For example, the transmission network and interconnections between systems may act as a flexibility backbone and as the 'net' balancer, for example in a largely integrated system with different climatic zones, such as in Europe. Flexible tariffs can sensitise customers to become aware of their energy use and encourage certain behaviours helpful to network operations. Decarbonisation in other sectors may create opportunities for further flexibility, in particular the electrification of transport and heating.

The extent to which these options are used will determine how quickly the system moves toward decentralisation, and bottom-up optimisation. For example, capacity payments in the United Kingdom provide financial support to the incumbent producers, and help to maintain the status quo. In Italy, a capacity market is envisaged, where producers will receive a remuneration for the generation capacity they make available. The European Commission approved the design of the Italian capacity market in February 2018. Capacity payments can significantly distort the market and offer financial support for a broad range of operators, often which other policy objectives are seeking to phase out (as has been the case in GB and its support for diesel) (Lockwood 2017). In Germany, the government created a requirement that network operators procure 2 GW of capacity to be held in reserve outside the market. This scheme not only bears advantages for incumbent generators, but might further restrict the balancing market to, for example, the detriment of demand side measures. Similarly, exporting excess power through large interconnections may reduce the need for much smaller scale storage and also negatively affect the economic case for demand response.

A different framework for the provision of grid services may better fit the requirements of the new system in terms of cost recovery, with every unit, including the low-carbon ones connected to medium and low voltage networks, being able to participate in network services if they wish to, thanks to the low-cost control technologies now available. For instance, in the wholesale electricity market in Australia the reserve capacity is undergoing a review for change to a capacity auction to commence in 2021, including changes to rules for demand side management, due to the current over-capacity and associated costs to consumers. Trialling has begun to alleviate demand peaks during high summer temperatures or to control the frequency of the network with virtual power plants (VPPs), composed of up to 1000 domestic solar and battery systems.

One policy option is to work towards a hierarchy of flexibility measures, with priorities for those with higher system efficiencies, higher greenhouse gas (GHG) reduction potentials, and those that have longer-term value. This is not a static consideration, especially given the system changes that large data management capabilities will bring, and the opportunity to evaluate and match supply and demand regionally or even locally.

This, plus an attempt to deliver greater integration across sectors, implies a move from the traditional top-down optimisation to a more bottom-up optimisation. Regulation may achieve superior outcomes if it is based on principle of subsidiarity, where first the total energy use is reduced on a local level – house by house, street by street – and then resources are integrated as much as possible towards the next higher level, before expensive high-tension networks and interconnectors are built. Government or the regulatory agency could entitle a Distribution System Provider or Distribution System Operator to prioritise incentives for the different mechanisms.

Any monetisation of flexibility services will lead to a shift in the allocation of revenues among economic agents. Thus, while evidence is now showing that providing more value in the energy system for flexible operations enables a more cost-effective whole system development and operation (i.e. cheaper overall and therefore beneficial for customers), there will be resistance to those changes (Shakoor et al. 2017). A balance has to be found between the new rules and incentives within those markets, networks, tariffs and services, and new institutions and actors – whether distribution market coordinators; system operators, and so on – who enable and coordinate the services and resources.

Strengthening the role of local system coordinators, thereby allowing for a coexistence of the central grid and decentralised micro-grids

In developing countries with fast growing electricity markets, such as India, the central grid suffers from low reliability and insufficient coverage. In this context, the construction of decentralised micro-grids could be encouraged and incentivised to complement the existing infrastructure and leapfrog towards more reliable, decentralised supply. In industrialised countries and states with a fully functional and reliable central grid, such as Australia or New York, decentralised micro-grids may be more resilient against climatic events like Hurricane Sandy in New York or storms and heat waves in Australia.

To accommodate more generation on the distribution network its system operator needs to have greater power and more strategic oversight. The DNO, which has often been relatively passive, with a fixed rate of return on their asset base, may be replaced by a coordinating distribution entity.

In New York State, distribution system providers (DSPs) encompass a new system function intended to coordinate an area system operation and to stimulate markets in that area. The distribution utilities retain both wires and system operator functions and have a public service obligation placed on them. The utility currently obtains its revenues from traditional cost of service. In the future, this is likely to move more towards performance-based regulation, one desired output of which would be a resource and cost-efficient system operation. This is effectively a new way of allocating costs in the energy system – more closely reflecting its use and needs, and the value of different resources within the energy system – including distributed energy resources, flexibility and demand side management.

New York state illuminates that regulation is adapting to the new configuration within energy systems. The state's system coordinators are still obliged to fulfil public policy goals. They are incentivised to minimise infrastructure costs of system transformation, and to maximise customer satisfaction. Tariffs become important to encouraging customer connection to their energy use, and this in turn links to network development (and network regulation) and meeting public goals. Currently network operators receive all the revenue related to networks, but over time new ways of paying for networks and other system functions may emerge – again more closely related to the system use, the providers of new services, and to what customers want the system, and networks, to do. Since liberalisation, electricity markets have always had certain links to network operation and system costs, but this is likely to become more complex, as greater levels of decentralisation and demand response occur. Thus, system operation, network charging, tariffs and markets – which have always worked together – are becoming more closely intertwined, and sophisticated. Australia is currently the world's most extreme example of this. On-site generation has reached a point where it makes financial sense for households to use solar and storage, even without subsidies. However, network charging and the regulatory mechanisms are lagging behind this momentum, and many systemic problems could have been avoided if they had been addressed earlier.

In the example from New York, system operation includes managing the distribution wires. But equally, a distribution market facilitator could be a 'system operator only' company or not-for-profit institution, while the distribution wires company becomes a regulated entity with a new role. Whatever configuration of the entity at distribution level, it will have to interact with the transmission system operators. A new system coordinator must allow and encourage the development of systems that support different functions on the grid, such as demand side actions, generation and storage. Small actors will often combine activities to accumulate revenues and in doing so offer important supply and grid stability services. A distribution system coordinator needs to recognise these advantages of multi-functionalism. It is still too early to know how these entities will be set up and how their interactions will work – both in developing and industrialised countries. It is clear, though, that the roles of both distribution and transmission companies will change.

Reforming regulation – including performance-based elements

Often, new business models develop *despite* the system rather than with the help of it. In many countries, the existing system maintains payments for the 'old' technologies and services and does not provide payments for the services that would enable the new, flexible system, for example many of capacity market incentives, which favour existing generators over demand side measures. Its institutions do not act as a driver, but as a barrier to emerging innovations.

Technology changes and the subsequent requirement for new operational regimes, coupled with emerging opportunities for active consumer engagement, are driving the need for far-reaching regulatory reforms. The new regulatory framework should be ambition driven, shaping the regulatory framework towards clearly defined policy objectives. If the energy policy remains too focused on conventional centralised technologies, change may be slower and, in the longer term, more expensive, due to larger stranded assets and wasted opportunities (Shakoor et al. 2017).

The traditional cost-of-service method of regulation requires utilities calculating their costs for the next X amount of years, the regulator checking their calculations and agreeing on the money they can spend over the time period. This amount is then turned into a charge on customers.

By contrast, performance-based regulation (PBR) is a form of regulation that aims to incentivise outputs in return for payments. It is very different from the more traditional cost-of-service mechanism. PBR decides what it wants to achieve (desired outputs) and then establishes an incentive mechanism whereby the utility is paid to the extent it delivers the desired outputs, as opposed to cost-of-service regulation. The value of outputs has to be worked out dynamically, as they will change over time, so that the payment to utilities per output is not too great or too low. Under the regime of performance-based regulation, inputs may change provided the desired outcomes are met, which means that there will be more flexibility of choice in delivering those outputs rather than being locked into the inputs. This regulatory regime is likely to lead to a better use of resources and cost-efficient system operation, as the report of the states of New York and California suggest. It also facilitates dynamically linking revenues, tariffs, connections and network operation charges with the desired market design.

Compared to cost-of-service regulation, PBR is also more flexible and better placed to incentivise these requirements from a public policy point of view, because outputs can be more easily changed. It also facilitates dynamically linking revenues, tariffs, connections and network operation charges with the desired market design.

Reforming regulation will encompass how to deal with winners and losers of the new regime. Trade-offs have to be discussed to ensure that, on the one side, networks are paid for and public service obligations are met, and on the other side the wishes of prosumers and users to have a high degree of autonomy are respected. In the move from cost-plus regulation to one where a higher proportion of network fees are linked to performance-based regulation, the roles and responsibilities of all stakeholders need to be addressed.

Reassessing investments in the long-distance transmission grid, given the rise of decentralised energy supply

As more customers connect to PV, they use fewer units of electricity, and the transmission and distribution lines must be paid for by fewer customers on fewer units, if the same governance for network regulation and for charging for network use is maintained. This increases the network portion of the bills and makes self-generation more economically attractive. This is the so-called 'death spiral' for the conventional energy supply industry. The reaction from some governments or regulatory agencies has been an attempt to stop subsidising the deployment of PV – rather than seeing it as part of a move to a sustainable energy system. Consequently, feed-in-tariffs for small-scale renewables have been reduced in an increasing number of policy frameworks, for example Germany, and an additional network charge 'or insurance premium' is proposed for consumers that self-generate, because they use – in conventional energy provision terms – the grid as backup (Gosden 2016).

In many countries, the construction of new and reinforcement of existing transmission infrastructure incurs costs in the billion-dollar range for final customers. These investments may lead to stranded assets, because the uptake of local supply reduces the need for long-distance transportation of the electricity. Some countries, such as Australia, have recognised the need for a reassessment of the necessity of long-haul transmission investments.

Network utilities have hitherto made their money primarily from their cost of service regulated payments; per unit of energy transmitted across their networks; and from connections to their grids. They have been in control of how their network is used and operated – who connects and how much those connections cost. Increasingly, however, as technologies decentralise there are new ways of ownership, network connection and network use. With these changes, the structure and origin of revenues for network entities have to be reassessed.

More decentralised production changes the volumes of electricity flowing across different segments of the grid. The sources, predictability and volumes on the transmission system will change as a result of new, sometimes large-scale renewable generation, such as offshore wind. By contrast, the tendency towards more production and consumption within the same regional distribution grid may reduce the overall flows in transmission systems. Self-production has already resulted in an increasing number of consumers who have reduced their consumption from the electricity grid, leading to a decline in the overall revenues for grid operators. Given that the grid operation costs are largely fixed, grid operators will, all other factors remaining, have to increase their unit cost per kWh of transported electricity. In turn, this encourages customers to buy more on-site generation, and raises important questions about how further grid costs need to be allocated to active and passive consumers alike.

In New York, the reduction in the revenue for utilities generated from electricity sales across the grid may be compensated for by payments for meeting specific policy objectives. In other countries, such as Australia, consideration is being given to a tariff similar to mobile phone charging, where the customer would choose a plan based on their peak kilowatt usage, for example not exceeding 3 kW of consumption at any time. If consumption is above this limit they would have to pay a fee. In Italy, an increasing part of the bill will be charged per unit of capacity and not just on consumption.

New rules should enable distributed generation while ensuring the network remains reliable and secure. In Italy, electricity generation is a liberalised activity, and grid operators are obliged to connect all renewable generators at a cost which is proportional to the distance from the connection point. However, as of 2018, the owner of a renewable energy plant does not have alternative solutions to self-consumption or sale to the grid. The direct sale of electricity to other consumers, as well as load aggregation, is not permitted, with the exception of the one-to-one supply under SEU ('Sistemi Efficienti di Utenza' or Efficient User System) scheme.

The next step in the operation of networks will be the ability of individual producers to sell directly to consumers. Blockchain and other open ledger technologies are now being tested in some countries – for example in Vienna with the municipal utility Wien Energie – and may accelerate changes in the regulation of the network: In a peer-to-peer scheme, one neighbour might want to connect to the grid and sell to another neighbour, but might want to pay only for use of a few metres of distribution grid. How should this be paid for, and how will this feed into the overall cost of running an energy system?

Box 2: To upgrade networks or not?

As the share of decentralised renewable energies rises, the default response of a traditional government or regulatory agency is to allow and promote investments in the distribution and transmission grid. In Germany, for example, around 1800 km of high-voltage transmission lines are under construction to transport the offshore wind energy produced in the North Sea and in the Baltic Sea to the country's industrial hubs in the South. As grid operations in most countries are still regulated, the costs for reinforcement are borne by all consumers via grid fees and levies. In Germany, the costs of grid services and concessions have been almost uninterruptedly rising from 1.02 €-cts per kilowatt hour in 2009 to 6.79 €-cts in 2018 for an average household with a consumption of 3500 kilowatt hours per year (BDEW 2018).

If energy systems with a larger share of intermittent, weather-dependent power sources continue to be operated in the same way, then reliability problems are likely to increase. Grid operators have to intervene more frequently to maintain the balance between supply and demand. This happens either by shutting down individual renewable energy plants if there is excess generation injected into the grid, or activating additional conventional capacity in case of excess demand.

For example, interventions in Germany's largest transmission grid operated by private company TenneT increased from fewer than 10 interventions per year in 2003 to almost 1000 interventions in 2014 (Weinreich 2016). The costs of grid interventions of Germany's four large transmission grid operators rose from €436 m in 2014 to €1130 m in 2015 and €848 m in 2016. The decline from 2015 to 2016 was caused by a lower intake of wind and solar energy in 2016, as well as optimised operations and redispatch of the grid operators, according to the German federal grid agency (ZfK 2017). In 2016, compensation for temporarily shutting down renewable energy installations amounted to more than €370 m (ibid.).

However, if a country starts to operate their electricity system differently, and adds cheaper flexibility resources, then these expensive networks upgrades are not required and reliability problems would occur less frequently, which keeps a cap on infrastructure cost increases.

In addition, transmission grid operators could start building expertise and a digital and technical infrastructure to cope with the new and more challenging system requirements. For example, German transmission grid operator 50 Hertz was able to reduce costs for and quantities of congestion management from 2015 to 2017 by 47 and 41 per cent, respectively, because of the optimisation of redispatch, as well as new transmission connections (Reinke 2018).

Regulators should be open to differing analyses of different scales of development when deciding on their regulated company agreements.

At root, the way we cost our energy systems and energy provision is changing. Renewable electricity is not yet ready for a flat rate system, just taking fixed costs into account, but in many industrialised countries the cost of generating electricity is less than half the cost of the retail price to customers – the remainder is related to network, system and environmental costs.[50]

[50] In the United Kingdom, it is about one third, see http://www.energy-uk.org.uk/customers/about-your-energy-bill/the-breakdown-of-an-energy-bill.html

As renewables are becoming considerably cheaper, the non-energy costs of energy provision become greater, and the focus will turn to how to pay for networks, system operation and the social and environmental costs of energy use rather than for energy itself. This is an entirely new focus of energy system economics. Network charges and access rules, which role prosumers play in the system, and what obligations and rights they have, is a new fault-line in energy regulation.

An integrative approach to sector regulation

The conventional energy system tended to have separate sector regulation, for example in electricity and gas, and they were top down optimised with few players. As the energy system decarbonises and decentralises, the convergence of heat, mobility and power on the distribution level requires coordinated regulatory instruments and actions. Regulators have to be flexible to changes and establish processes whereby regulation can keep up with and be adaptive to changes – rather than undermining them.

The decarbonisation of the heat and cooling and the transport sector has not been as rapid as for power. Consequently, in addition to the promotion of the greater use of renewables in these sectors, for example biofuels in transport and district heating, the electrification of these sectors is being promoted to reduce emissions. In both contexts increased attention is being placed upon sector coupling, that enables the co-production, combined use and substitution of different supply and demand options. In addition, sector coupling may increase the resilience of the system, given the variability of renewable energy production.

With a Bloomberg New Energy Finance forecasts suggesting that there could be 11 million EV sales per year globally by 2025 (up from 1.1 million in 2017), smart charging of electric vehicles could massively expand and create an unprecedented opportunity for grid balancing through customer engagement (BNEF 2018). In some countries, such as Denmark, piloting has already begun (Fuelincluded 2017).

From a regulatory perspective, sector coupling raises important questions of how to set up a framework that does not only optimise the deployment of different technologies and distributed energy resources in individual sectors, but also for an encompassing regulatory and institutional framework. The electrification of new sectors is likely to increase electricity demand, which would be significant in many industrialised countries that have experienced no or little growth over the last decades.

Going forward, the development and implementation of a smart, holistic energy system will require coordination between the variable renewable producers, the transmission system operator (TSO) and distributed system provider (DSPs), the municipalities and even the vehicle fleet owners to achieve maximum efficiency and stability.

Most energy systems in the world have a very clear public service obligation on monopoly providers of services to customers, and they have customer licenses of some sort on the non-monopoly providers for other functions.

The changing energy world is altering the roles of different actors and stakeholders, but there still needs to be a clear requirement on actors and stakeholders to provide a certain level of service to customers. It is remarkable that the Public Service Commission (PSC) of New York, the Regulator, has come to the view that upholding their public service mandate, which is about a century old, can only be fulfilled by fundamental changes to their energy system – but they still hold fast to their mandate. Vulnerable customers will need to be looked after, and networks, if they are needed, still need to be paid for.

2.9.2 The way forward: transformation and acceleration

Energy systems are changing and becoming more decentralised for all the reasons for all the reasons and drivers discussed above this transformation needs to be undertaken in the most cost-effective way possible if it is to be accelerated with a parallel acceleration in greenhouse gas reduction. This chapter argues, from evidence taken from the country sections, that a key enabler of an accelerated transformation is a coordinating governance framework made up of 8 key elements, we now move to business models which can thrive from in situations where those governance mechanisms are in place.

2.9.3 References

BNEF. (2018), *Runaway 53GW solar boom in China pushed global clean energy investment ahead in 2017. Bloomberg New Energy Finance, 16 January 2018.* Retrieved February 15, 2019, from https://about.bnef.com/blog/runaway-53gw-solar-boom-in-china-pushed-global-clean-energy-investment-ahead-in-2017/

Burger, C., Trbovich, A. & Weinmann, J. (2018), *Vulnerabilities in smart meter infrastructure – can blockchain provide a solution? Results from a panel discussion at EventHorizon2017.* Berlin: German Energy Agency dena/ESMT. Retrieved from https://press.esmt.org/all-press-releases/blockchain-can-improve-data-security-energy-infrastructure

Butenko, A. (2018), *Active customers, aggregators and local energy communities in the proposed fourth electricity directive.* Retrieved February 14, 2019, from www.ogel.org/article.asp?key=3734

European Commission. (2016), Communication from the Commission to the European Parliament, the Council, the European Economic and Social Committee, the Committee of the Regions and the European Investment Bank, Clean Energy For All Europeans. 30 November 2016, COM (2016) 860 final.

European Commission. (2017), *Smart meter deployment in the European Union*. Retrieved October 16, 2017, from http://ses.jrc.ec.europa.eu/smart-metering-deployment-european-union

Finkel, A., Moses, K., Munro, C., Effeney, T. & O'Kane, M. (2017), *Independent review into the future security of the national electricity market – blueprint for the future. Canberra*. Retrieved August 5, 2017, from http://www.environment.gov.au/system/files/resources/1d6b0464-6162-4223-ac08-3395a6b1c7fa/files/electricity-market-review-final-report.pdf

Gosden, E. (2016), Households could be charged annual 'insurance premium' for access to electricity grid. *Daily Telegraph*. Retrieved July 5, 2016, from http://www.telegraph.co.uk/news/2016/05/29/households-could-be-charged-annual-insurance-premium-for-access/

Hoggett, R. (2016), *Rethinking the role of consumers in our evolving energy system, IGOV programme, University of Exeter, 8 July 2016*. Retrieved January 30, 2018, from http://projects.exeter.ac.uk/igov/new-thinking-the-changing-role-of-consumers-in-the-energy-system/

IETA. (2015), *Overlapping policies with the EU ETS, International Emissions Trading Association*. Retrieved November 20, 2016, from http://www.ieta.org/resources/EU/IETA_overlapping_policies_with_the_EU_ETA.pdf

REN21. (2017), Renewables 2017 global status report. Paris: REN21 Secretariat.

Sandbag. (2018), *The European power sector in 2017, state of affairs and review of current development analysis. Agora and Sandbag, January 2018*. Retrieved February 15, 2019, from https://sandbag.org.uk/wp-content/uploads/2018/01/EU-power-sector-report-2017.pdf

Shakoor, A., Davies, G., Strbac, G., Pudjianto, D., Teng, F., Papadaskapopoulos, D. & Aunedi., M. (2017), *Roadmap for flexibility services to 2030, a report to the Committee on Climate Change. Imperial College, May 2017*. Retrieved February 15, 2019, from https://www.theccc.org.uk/wp-content/uploads/2017/06/Roadmap-for-flexibility-services-to-2030-Poyry-and-Imperial-College-London.pdf

Weinreich, V. (2016), *Sicherheit der Elektroenergieversorgung im Zeichen der Energiewende, Entwicklung der Netzeingriffe in der TenneT-Regelzone seit 2003*. Retrieved February 15, 2019, from https://www.vde-kassel.de/resource/blob/692944/c6beec0a3555509f49fdc8b8322b048c/vortragsfolien-download-data.pdf

US Department of Energy. (2017), U.S. energy and employment report. Washington, DC: U.S. Department of Energy.

ZfK. (2017), *Zahl der Netzeingriffe ist 2016 deutlich gesunken*. Retrieved February 19, 2019, from https://www.zfk.de/politik/deutschland/artikel/zahl-der-netzeingriffe-ist-2016-deutlich-gesunken-2017-05-31/

CHAPTER 3

Business models beyond subsidies – which core competencies are needed?

Christoph Burger and Jens Weinmann

3.1 Energiewende 1.0 – 3.0: matching phases of energy transition and business models

As outlined in the introductory chapter, the energy transition is fuelled by two main drivers, which are reflected in the two main parts of this book.

The *top-down* drivers are typically governments or states – or regulatory institutions within these territorial administrative entities – that are striving for a reduction in greenhouse gases or want to promote certain industries and their technologies. Examples of countries, their policies and governance structures have been presented in the previous chapter.

In contrast, this chapter of the book is dedicated to the *bottom-up* drivers: individual businesses and start-ups that exploit niches within the new energy system to generate revenues by creating or joining new markets and platforms. They are emerging alongside the established players in the energy sector, in particular the electric utilities.

Liberalisation of the energy sector is generally the pre-condition for allowing these players to emerge. Liberalisation (of certain parts of the value chain) creates opportunities for trading electricity and natural gas on wholesale markets,

How to cite this book chapter:
Burger, C. and Weinmann, J. 2020. Business models beyond subsidies – which core competencies are needed?. In: Burger, C., Froggatt, A., Mitchell, C. and Weinmann, J. (eds.) *Decentralised Energy — a Global Game Changer.* Pp. 177–260. London: Ubiquity Press. DOI: https://doi.org/10.5334/bcf.k.

with all types of financial instruments being offered by multiple players. However, a bottom-up movement is also possible in country settings that still favour a fully regulated configuration of the electricity supply. For example, governments can also demand/require that a certain share of a load-serving entity's energy supply is derived from renewable sources (for a discussion, see also Kieffer & Couture 2015). These utilities could then issue tenders to other players and outsource the ramp-up of renewable capacity, with most closely following the existing model of independent power producers (IPPs), which supply certain amounts of energy, typically in long-term contracts with utilities or the government. This model of transformation functions via command-and-control rather than market-based mechanisms. California versus New York state would be classical examples of these two differing approaches.

Whereas regulatory policies, implementation, and rollouts may differ after the electricity market is liberalised, the move towards decentralisation typically encompasses three phases, which we call Energiewende 1.0, 2.0, and 3.0, or Phase I, Phase II, and Phase III (see Burger & Weinmann 2017 in the *Harvard Business Review* for further details). In technical terms, they can roughly be associated with a deployment of 'new' renewable energies with a supply share of less than 10 per cent in Phase I, when renewables still represent a niche of the electricity supply. In Phase II, their contribution rises to 10 to 40 per cent, and they become a major player in the supply portfolio. Phase III is characterised by an aggregate renewable-energy supply of more than 40 per cent, when they can be characterised as a dominant player by an aggregate contribution of more than 40 per cent and renewable energies as the dominant player (Baumgartner 2017).

This categorisation is only valid for the 'new' renewable-energy sources, in particular wind, solar, and biomass. For example, the United Kingdom meets some of its renewable targets by co-firing large, old coal stations with biomass. By contrast, countries with a high share of large-scale hydropower, such as Brazil, Norway, and Paraguay, do not enter this classification as Phase III countries – despite the fact that renewable energies play a dominant role in electricity supply – because the supply structure is based on centralised operations and control of the assets.[51]

Each of the three phases brings its own opportunities and challenges for policy makers as well as corporate players such as utilities and start-ups, as will be discussed in greater detail in Chapter 4.

---

[51] Hydropower has been in use for more than a century to produce electricity, and it does not feature the decentralised geographical pattern of 'new' renewable energies. In terms of structural similarities, in particular with regards to the financing and complexity of construction, offshore wind farms could be compared most closely to a conventional hydropower dam: Both require larger players with a strong financial endowment or backing by investors to provide the high levels of investment up front.

- **Phase I (Energiewende 1.0):** in Phase I of the energy transformation, countries explore opportunities to incentivise the deployment of renewable (non-hydro) energy sources. Start-ups benefit from public funding for the rollout and provide services. Owners of residential rooftop photovoltaic (PV), bioenergy villages, and also energy associations that operate wind turbines are at the heart of a 'civic power' movement, in which assets are owned by private individuals, without a utility being involved in the operation of the plants. In parallel, small and medium-sized enterprises (SMEs) and some power companies are investing in and deploying renewables.
- **Phase II (Energiewende 2.0):** Phase II of the energy transformation is characterised by civic power emerging as a third force in the market, complementing electric utilities and corporate new entrants in the energy sector. Civic power is one of the reasons why the decentralised energy revolution may become a 'global game changer' – it shifts the responsibility of electricity supply back to the citizens of a country.

  During Phase II, platforms are created that coordinate supply and demand or offer services such as aggregating existing capacity or loads (for peak shaving). When multiple atomised actors are involved, data has to be gathered and analysed. Individual owners of assets may not be sufficiently knowledgeable to deal with the complexities of the energy system, and they outsource that expertise to companies that specialise in providing these services.

  Regulators start modifying the initial incentive systems in Phase II. A continuation of directly or indirectly subsidising renewable energies is often countered by public opposition because of the heavy burden for ratepayers.

  Utilities adapt to the new market environment by reorganising their activities, as it happened for example in Germany, when incumbent utilities E.ON and RWE split their traditional thermal generation and trading units from their distribution, renewables and service-oriented business lines, and later merged these two new entities.
- **Phase III (Energiewende 3.0):** Phase III of the energy transformation, or 'Energiewende 3.0', has yet to be seen in any country. Non-hydro renewable energies will become the major player in the supply structure. The marginal costs of renewable energies such as wind and solar are practically zero, so there will be a resource abundance at certain times, and extreme scarcity (and high prices) at other times. Hence, storage will become a major issue for policy makers.

  The electricity supply industry will be forced to leave its roots as public infrastructure service and transform into truly private businesses, offering customised solutions for each consumer, while independent system operators or private transaction platforms take over responsibilities of grid control. As with many other aspects of our lives, energy may become an individualised choice, with each of us determining the amount of risk we are willingness to take. The customer proposition may become central in this phase but it will not necessarily mean that customers will become active beyond choosing a proposition they like. Energy will, of course, remain a public infrastructure service, but the state's involvement will shrink, as happened in the telecommunications and aviation sectors.

3.2 Start-ups pave the way towards a new energy system

Compared to the 'safety and reliability' paradigm that has dominated the mindset of executives at electric utilities until very recently, start-ups typically have a different mentality and culture. A willingness to take risks and to 'pivot' – which means a radical change in the business model – are essential ingredients in their trajectory towards success. Large companies have started complementing their traditional research and development units with external input to increase diversity and the spectrum of potential future business options. They acquire smaller companies with new ideas, for example Centrica in Great Britain, participate in accelerators, establish incubators and venture capital funds, foster a culture of intrapreneurship, that is, internal entrepreneurship, and are (slowly) changing their business models – from selling a commodity to becoming a provider of integrated service solutions.

As in many other industries, the electricity supply industry benefits from innovations and unconventional ideas being brought to the market by start-ups, founders, and entrepreneurs. The focus of the interviews used in this part is, hence, on the insights generated by their experiences.

The start-ups with whom the interviews were conducted have been selected to mirror the changes that are taking place in both industrialised and developing countries. Start-ups in developing countries may have completely different business models than start-ups in industrialised countries. This is because the former use decentralised energy generation to *complement* the existing grid infrastructure in areas where grid connections are not established, whereas decentralised energy in industrialised countries is a *substitute* and replacement of existing supply (and respective institutional) structures.

Furthermore, they have been chosen as representative examples of the three phases of the energy transformation described above.

One case study of business models during Phase I of the energy transformation is Envio Systems (no. 1), a Canadian start-up that offers a low-cost solution to enhance the energy efficiency of existing commercial buildings. Moving towards a partially autonomous system with a larger share of renewable, intermittent energy, Timo Leukefeld (no. 2) has developed a commercial solution for an (almost) energy-autonomous house, whereas Entelios (no. 3) has been one of the first providers of demand response services in Europe. For most industrialised countries, Phase III of the energy transformation is a vision, a future perspective, but in the context of isolated, rural areas in the developing world it is factual reality. The business models of the three start-ups SOLshare (no. 4), Mobisol (no. 5) and Solarkiosk (no. 6) offer alternatives to the connection to

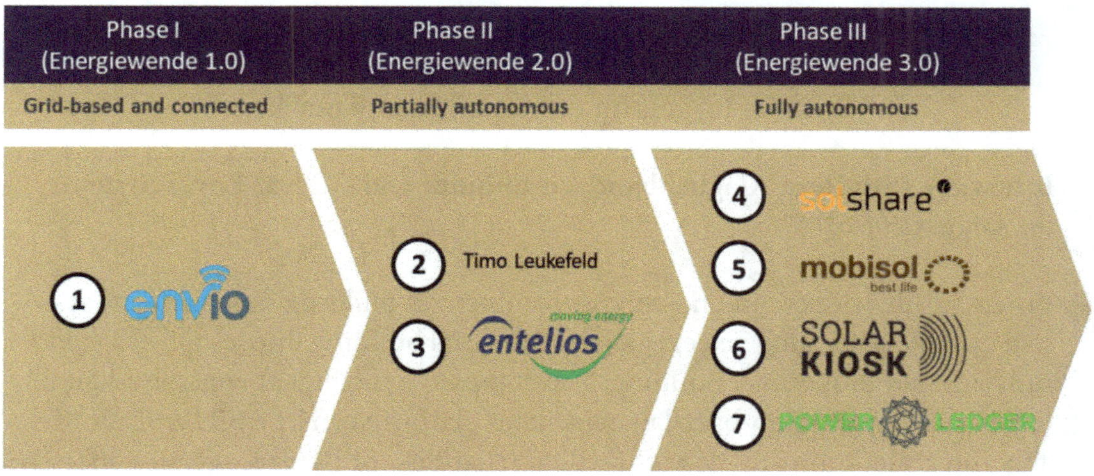

Figure 20: Classification of interviews according to phases of the energy transformation.

Source: Authors' contribution.

the central grid. By contrast, Australian start-up Power Ledger (no. 7) is a pioneer in peer-to-peer trading and Blockchain-based decentralisation of power supply in highly developed urban and suburban settings.

Innovators test new business models. Only a few survive, others inspire new ones. Scaling becomes important, and this might be a reason why many disappear. Since some of the interviews have been conducted, substantial changes in the business models or career paths of the interviewees have taken place. Some of the startups have filed for insolvency, others have been acquired by larger competitors. All interviews must therefore be interpreted as snapshots of their situations of the world at a certain point in time. Despite these developments, the interviewees kindly agreed to authorise the publication of the interviews after providing updates on their ventures before the manuscript was completed and submitted to the publisher.

3.2.1 References

Baumgartner, D. (2017), How to integrate a high share of renewables into the grid? HEC-ESMT Energy Course. C. Burger and J. Weinmann. Berlin, Elia Grid International.

Burger, C. & Weinmann, J. (2017), The 3 stages of a country embracing renewable energy, *Harvard Business Review*. Retrieved February 11, 2019, from https://hbr.org/2017/04/the-3-stages-of-a-country-embracing-renewable-energy, accessed 11 February 2019.

Kieffer, G. & Couture, T. D. (2015), Renewable energy target setting. Abu Dhabi: IRENA.

3.3 Envio Systems: redefining building efficiency – Envio Systems targets an untapped legacy market

Interview with Reza Alaghehband, co-founder and CEO at Envio Systems, on August 26, 2016

Envio has developed an end-to-end commercial building management system capable of turning any existing commercial building into a sophisticated, fully autonomous Smart Building. The company's intelligent controls, digital infrastructure and web-based management platform can be easily and affordably integrated into any type of facility, regardless of its size, age, or sophistication level. According to the start-up, this is a breakthrough solution that affordably enables the AI (artificial intelligence)-powered, fully autonomous management of commercial buildings without the need to replace any existing infrastructure. It pays for itself from the energy savings in less than three years, which is a fraction of the cost of existing solutions. The systems create smarter facilities that operate using 20–70 per cent less energy (Envio Systems 2018).

3.3.1 Technology and business model

What are the barriers preventing 87 per cent of all commercial buildings from adopting advanced automation systems? The answer is simple: the costs heavily outweigh the benefits. We have developed our system as solution to this question.

With our system we wanted to overcome three major challenges: First, how can we eliminate structural complexity, which would require a high level of expertise to operate or troubleshoot? We decided to supply a pre-wired, easy-to-deploy set of systems ready for installation. Any electrician globally can install and test it using our simple app. It reduces the high cost of installation by enabling a wireless plug and play set-up. The second challenge was compatibility: How do we make it interoperable with the vast majority of systems? We developed adaptive and remotely configurable hardware flexible enough for every scenario. Third, how can we minimise customisation and circumvent an advanced configuration required for each facility? We achieve this with self-programming, commissioning, and learning devices configured via templates over the web.

Technology

The system we have developed is comprised of three key components. The first is the 'Cube', a plug and play IoT (Internet of Things) controller that universally connects to any legacy device in a building (light switches, thermostats, boilers, chillers, fans, pumps, valves, metres) to enhance them with bi-directional communication, web connectivity, and enable building-wide interoperability.

The second component is the Envio Gateway, which is a translator between communication protocols and legacy systems as well as a bridge between the Cubes to the cloud and back.

The third is our web-based platform, which collects, analyses, and visually displays all components and also hosts our algorithms. This is the key component, in that it utilises real-time and historic information to determine the optimal setting for every single component, building-wide. Then it sends the settings back to the building for each Cube to execute.

How can we consolidate proven energy-efficiency functionalities that every building needs while preparing them for the future? Where can we create the highest dollar-for-dollar value? How can we integrate everything into one device? Instead of having redundancy in the hardware of our buildings, which increases costs and makes them less financially viable, we wanted to combine them and use one device with the advancements in processing power to collect all that information. We wanted to be able to make active decisions and communicate with various building systems using the existing infrastructure. Rather than ripping things out and putting new equipment in, we simply integrate the device with the existing equipment, making it smarter and helping it to operate more efficiently.

Commercial buildings operate similarly to factories, ensuring that the people inside are comfortable by managing climate, lighting, and ventilation. Factories are only efficient when fine-tuned for consistency. The difficulty is that, in commercial buildings, there are dozens of completely unpredictable variables such as occupancy, outdoor temperature, heat loss, sunlight, and efficiency losses, which are unique to each building and constantly change. Using machine learning, our technology automatically learns and adapts to each building.

Our core IP is built right into the intelligence. Based on the sensor information that is located around the area, the systems decide on what is being told to the radiators, the fans, etc., unless we give them other instructions based on additional information coming from all the sensors we have in the area. When rooms are occupied, in most cases we allow people to have full control over what is going on, depending on how they like their environment. But once they leave, the majority of the systems will continue to heat and light that room and have the same level of ventilation entering the building and the room. We prevent that, and that is how we go about saving energy.

One of the first things we do is to implement our system within the building. With every Cube that we equip, we have about six sensors to monitor that area. In every room with a Cube, we are able to track the rate of heat loss and gain,

Figure 21: Envio Systems' Cube.
Source: Envio Systems (2017).

the efficiency of the fans, and the ventilation and heating system. We are able to tell what strategies and products would be most beneficial in which areas, and where they should target first. It is never a shotgun approach to put everything in because not every installation is going to give you the returns that you need. Our system diagnoses where the most energy is being lost and why. Using that information, one can make much more intelligent decisions. We recommend to the building owners and operators the best dollar-for-dollar investment.

Our Cubes are also equipped with a CO_2 sensor for demand-control ventilation. That is probably one of the highest cost components within our system for each system that we implement, but it pays for itself within that period, so it is extremely beneficial to have.

We are very satisfied with the functionality and the payback of our system. In most scenarios, we are offering a cheaper alternative to what our customers would normally be implementing. Energy conservation and efficiency in commercial, industrial, and even residential buildings is a very straight-forward and repeatable process. First and foremost, you need to measure and understand what is going on in there before you can even make any decisions. If you are just looking at a bill, you do not know whether the energy consumption primarily came at the beginning of the month, at the end of the month, or even if it is an annual total. Time-based information is needed to understand what is going on within buildings. Once you have that information, you can prioritise which features you want to implement. Many companies come in with technologies that just sense information.

For me, that is a giant waste of money, because sensing information is great. But I have found in 99 per cent of the cases that, unless there is a major flaw

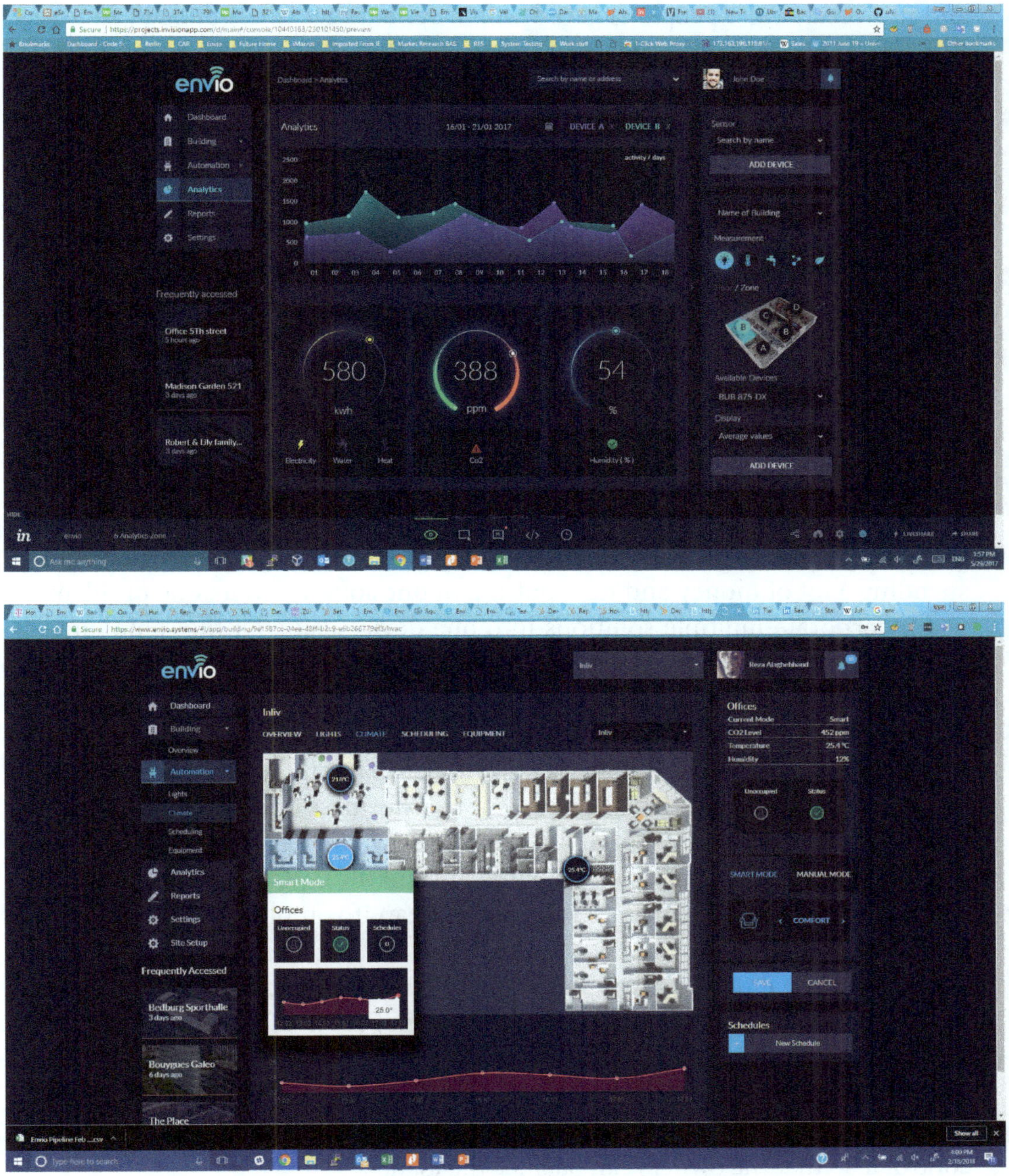

Figure 22: Envio Systems' BASE dashboard.
Source: Envio Systems (2017).

in the operations that they are completely missing, nobody is going to go and make the small adjustments that are continuously needed. The building may not have the capability for someone to make those small adjustments. But if you were to combine and utilise existing infrastructure and take the information and do something with it autonomously, it would actually require someone to take that part on. It brings people on board in a more active way because everybody loves the technology: It is fascinating to see everything going on, and it is great to get people engaged with a system that is not difficult to use. Bringing together various parts and various functionalities under one platform gives users the opportunity to have something like an operating system for their building and an incredibly valuable asset.

One should first install the energy metering system along with sensors and controllers. This is because there are basic functionalities that buildings older than 20 years do not have, which we implement in order for them to operate more energy-efficiently. Those are a required package. The energy monitoring system needs to be incorporated, otherwise we do not know what effect, say, turning down the fans is having in real time. In our package, we encompass energy monitoring. We know where the baseline is for the average commercial building, and we know the energy intensity per square foot/metre that should be used within these buildings. If a building's usage levels are above that, one knows that something is wrong. Around 40 per cent of energy in buildings is wasted, somewhat analogous to marketing: 50 per cent of your marketing budget is wasted; the problem is to find that 50 per cent. Instead of spending a lot of money and time putting in one solution to fix one problem, you incorporate multiple solutions into one device using all the same intelligence and share that intelligence with all the other rooms, then you have a synchronous system that looks not just at one part of the building but at the building as a whole. Implementation costs are lower because of the savings they produce. It would not make economic sense to buy separate sensors for the lights and HVAC if I installed current transformers (CTs) to measure and monitor electricity consumption. None of that would get a payback for you because of the redundancies in the hardware and the costs for manufacturing every single device. Sensors are incredibly cheap if you plug all of them into one device.

If consumers allow for active demand-side management, they will get benefits from it. But it also opens up a market that they are not fully in yet and have not yet figured out how to penetrate.

Our system does not need to be connected to the grid. If a building has a decentralised energy system, one of the most valuable features is being able to control it so that the owners can balance the load.

It is not a challenge to create and develop a system like ours. The challenge rather is backward compatibility with existing systems and understanding how to integrate them. We have a competitive edge in analysing all the information and having the building adjust to operating in the most energy-efficient way. It is no longer about simple control strategies. We are able to detect how much energy is being used by fans and the ventilation system, and we can control the pumps and the hot water boiler temperature. Over time, all of these systems become inefficient at different rates. We can determine the most efficient solution – increasing the speed of the fan or adjusting the hot water temperature – by applying intuitive and real-time operations and adjustments.

Business model

There is an opportunity for savings in every single building. In trying to make the solution viable, one has to determine the potential for reducing energy. If you want it to happen during a certain payback period, what would be the desired pricing model that fits with the reductions? The biggest component of selling energy efficiency is really the pricing model, understanding energy costs and prices, looking at the energy statistics and analysis, and generating energy savings.

We started in one of the markets with the lowest energy prices, almost globally, which was Calgary, Alberta, in Canada. The concept was that if we succeeded in making our system economically viable there, we could make it viable anywhere in the world. Alberta is not a very progressive province in terms of utility rebates, so we did not even have that luxury to fall back on if we made a business case. The cost savings had to be strictly from the energy savings that we produced.

Having worked for years within the industry, I know that there is a finite payback period of the systems and solutions implemented in terms of energy efficiency and conservation. We must meet that threshold. Any measure above that threshold may be realised, but it goes towards next year's budgeting, or the year after, which creates a much longer sales cycle. That happens to be one of the killers of the existing business model. There is no chance for generating revenue to finance business operations so fast.

If you are targeting the B2B market, no matter how sexy it looks, the bottom line is what matters for them. As a general rule of thumb, investments should pay off over a period of 36 months. Sometimes it is less, but if it does not provide some payback within 36 months, it is usually a very, very difficult proposition to sell. You have to figure out the savings and work backwards to figure out your pricing model.

Simplicity is the key to our business model. I have been working on this business model for about six years now. It has evolved over time and adapted. The original business model was a shared savings approach. It failed because nobody would touch it. There were tremendous measurement issues. We thought our system would overcome that because we were putting sensors and measurement devices everywhere to detect exactly how much we were saving. However, what happens generally is that one has no control over what the customers install in their buildings over time, and it becomes too complicated.

We are currently conflicted about our pricing model, and we may go in a different direction. Together with a very large utility partner, we are creating something incredibly disruptive for the market. Instead of us charging for the controllers, we are looking at hardware-as-a-service model, whereby clients simply pay on a monthly basis for us to reduce their energy costs.

We normally have recurring revenue. We can reduce the cost of the hardware to a minimum. If we increase the software service revenue to about 75 per cent of the savings we are generating, we can provide the hardware free of charge to rapidly capture market share. We can go to clients and offer them hardware and software at little to no cost, with no initial capital investment required. It will help them to reduce energy costs without any upfront costs and helps to resolve the value dilemma – the owner not being the operator.

The partnering utility derives benefits from being able to manage buildings more intelligently and to communicate with the interior devices of the building, bringing all these systems together and connecting them.

We do our pricing model by market. In Canada we have much lower pricing. We do it in 500 Canadian dollars because the utility rates and energy rates are much lower there. We increase our pricing model in markets where we have the opportunity for, say, utility rebates to increase gross margins.

We charge €15 per controller per month. On the platform, our customers can monitor how much energy they save. For example, if customer spend US$18,000 per month for our controllers, they may save US$25,000 per month in energy costs.

We would offer this service at a lower rate in places such as Calgary, but it would be lower on our list of preferred markets. In Canada, we would first target the province of British Columbia, where the government offers a rebate of 50 per cent.

3.3.2 History and organisation

We are currently employing about 14 people now. We add roughly one or two people per month. Realistically, within the next 12 months, we will need to have a total of 35–40 people. We retain our employees by having them be invested in the company so that it is not so easy to lure them away. The majority of our new hires will be working in the area of engineering and research and development. We will change the industry with our next product.

Our company will be set up in a decentralised way. If we are too centralised, we cannot react and manage appropriately. Our servers will be located wherever we are on that continent. Due to legislation, we have to make sure that our European servers are within Europe. As our data is highly secretive, it has to be secured. We will most likely have our centralised R&D facilities in Berlin. Our goal is to create an entire platform that other hardware systems can integrate into. We want to overcome the problem of everybody designing their own user interface, and thereby overwhelming clients.

3.3.3 Scaling and cooperations

Scaling

We have developed the first Building Automation technology, which is designed for rapid global scalability. Within the building controls industry, you will find that most companies are regional except for the big four (Siemens, Johnson Controls, Honeywell, and Schneider Electric). This is due to the requirement of having specialised installers, who must go on site to install, configure, and commission the systems. Our solution can be installed by the building owners' existing electrical contractors, and the set-up and configuration is either automated or completed remotely. We continuously monitor each sensor and controller for failures. In case of such an event, they can quickly be swapped out and will automatically download the previous unit's programming within seconds.

Our focus is on where we can have the most impact in terms of profitability, so we target the key markets first, the ones with higher energy rates. In Europe, electricity rates are significantly higher than in Canada. California, New York, and almost any place on the US east coast is available because there are fewer natural resources. As people move more towards renewable energies, energy costs become significantly higher.

There are tremendous opportunities available in developing countries because scarcity creates an increase in price, and most of those countries are just deploying diesel generators to power things, which is much more expensive than the costs for transmission lines and electricity here. Our system would allow for a more streamlined management of electricity grids, which would prevent blackouts and brownouts. Utilities could manage the grid better by simply turning down devices, which is a more granular control of energy consumption.

Cooperations

I recognised that we needed a partner a long time ago. The challenge was having people understand our business model, understand the industry, and say: 'Wow – that is very different from what is out there!' Except for those entrenched in the industry, people were not able to differentiate how we were different from Nest. It took a very long time to overcome that, and explaining and clarifying our value proposition was a challenge. We have a very long time to refine our business model. Gridpoint essentially bought three companies and tried to glue them together to create a product. It is a beautiful business model, except when overhead costs are so high that a company has to borrow hundreds of millions of dollars to keep operations going. Gridpoint did not realise how slowly the scaling process advances. As they move forward, they need to bring on the right team members and go to the right market. Unfortunately, they got to a point where investors became a little nervous about how deep they were going, so they were purchased.

The leading utilities know that they are behemoths and that they move incredibly slowly. They know they have to disrupt their business model before a company like us comes and does it before them. They have to be proactive rather than reactive. Major utilities understand that their business model is going to be dead probably within the next 10 to 20 years. It will have a significant impact, with big power stations just closing down due to government mandates, as well as decentralised energy taking a huge chunk of revenue away from them. They need to adjust their business model in order to stay alive.

Our biggest hurdle will be distribution. The utility would help us with distribution as well as logistics and operations. One of their biggest assets is information. They know how to scale and to deploy infrastructure almost better than any existing company. They know the market and who the largest energy users are. They can actually access their data and verify which buildings are most likely to be energy-inefficient and determine where to go with that information and how they can help them. If we establish a partnership with them, they can bring on the manufacturing side that we need support with and they can help us with the marketing and branding – they know the industry. They can provide some of their best engineers to us so that we can find out how to really take advantage of the data and the implications of that. They can provide us with resources that we cannot acquire fast enough to scale as needed.

We are also in talks with an elevator company. The reason why we would like to cooperate with them is that we would like to have the resources to service our clients. We would work with them as operators of our devices, whereas the cooperation with the utility would be mainly focussed on sales.

Most owners own more than buildings and have diverse portfolios. There are companies that own dozens, if not hundreds of buildings. We are currently working with a major property management company that is in a public–private partnership. Their role is to retrofit all city buildings to make them more energy-efficient to achieve the 2020 goals. Our first project together with them was a school, and it has already been implemented.

We are continuously looking for partners with whom we can easily bring onto a very sophisticated interface and exchange information between various systems to help the owners operate their buildings more efficiently. We are looking for apps. We are looking for solar companies or storage companies as partners. We want to help in getting devices to hit the market faster. It is a long process to establish a product in the market. With our support, we can direct these companies to the right clients.

3.3.4 Market outlook and competitive environment

We are different from other providers such as Nest, which produces devices for residential applications. When you do a residential application in one location, you are missing out on all the information around you. You are making a decision for maybe not only one specific room, but for the entire living area – upstairs, downstairs – which is a terrible way of going about it because you do not have sample data from other areas. Every single one of our Cubes is placed in various rooms to control that zone and that area, based on the information it is getting rather than assuming that the one single room with the sensor is the norm within that building.

When we work in commercial buildings, we usually have access to the ventilation system and ventilation controls, which provides a huge savings opportunity for us, just with being able to adjust ventilation based on the amount of carbon dioxide in there. Nest's focus is not energy savings, but rather it is a business strategy. They are not selling their hardware based on the fact that their customers will save a certain fixed amount. Rather, they sell it to give customers control – it gives you a little bit of information, and it looks very sexy. It is a consumer product.

3.3.5 Interviewee biography

Reza Alaghehband is a serial entrepreneur originally from Vancouver, British Columbia, now residing in Calgary, Alberta. He enjoys the challenge of disrupting the status quo through innovation and unique strategies. He possesses a Bachelor of Commerce in Entrepreneurial Expertise from Royal Roads University. Reza has been a featured presenter internationally at universities, industry conferences, and multinational organisations discussing technology, entrepreneurship, and innovation. Reza was the youngest member to be accepted into the Service Core of Retired Executives (SCORE) in the organisation's history.

He has been successfully founding companies in multiple industries, focussing primarily on materials engineering, renewable energy, and most recently energy conservation as the founder and CEO of award-winning Envio Systems, Inc.

3.3.6 References

Envio Systems. (2018), *Smart buildings simplified*. Retrieved February 10, 2018, from https://enviosystems.com/, accessed 10 February 2018.

3.4 Timo Leukefeld: a business model for an energy-autonomous house without subsidies

Interview with Prof. Timo Leukefeld, TU Freiberg, on April 6, 2017

Can a business model based on energy autonomy also be viable for industrialised countries? Germany, for example, has one of the most reliable electricity systems in the world, with the System Average Interruption Duration Index (SAIDI) typically hovering around two minutes per year, and a 100 per cent electrification rate. Yet, Timo Leukefeld, lecturer at the Technical University Bergakademie Freiberg in the German state of Saxony, is promoting an energy-autonomous house that produces its own heat and power – including the electricity for an electric car – with solar power.

3.4.1 Technology

I am a lecturer at the Technical University Bergakademie Freiberg. Together with my team, I have developed a financially viable and commercially successful concept for an energy-autonomous house.

I live and work in an energy-autonomous house that offers around 160 square metres with a solar thermal rooftop collector of 46 sq. m), photovoltaic (PV) panels measuring 58 sq. m), a 9300 litre long-term water storage tank, and an electric battery with a capacity of 56 kWh, based on lead technology. My consulting practice and I were in charge of the design of the energy-related installations.

Staff from the Technical University of Freiberg examined the energy performance of the building over the last four years with 190 sensors and observed that the house reached almost 100 per cent electric energy autonomy, including the supply for charging an electric vehicle. In thermal supply for heating and warm water, 70 per cent autonomy was achieved. The remaining 30 per cent of additional heating requirements could be met by using 3 cubic meters of fuelwood in the winter, which led to a carbon-neutral energy balance all year around.

The outer protection layer of my buildings fulfils similar quality standards as a so-called KfW 55 house, which uses 55 per cent less primary energy than a standard house. With the solar thermal system and PV panels, energy costs of a prototypical passive house can be further reduced – by more than two-thirds. The electric vehicle can be charged 10 to 11 months per year with locally produced electricity. Only in December and January must additional electricity be purchased from the energy supply company to charge the car's battery. These buildings achieve very high levels of energy autonomy, and the costs for

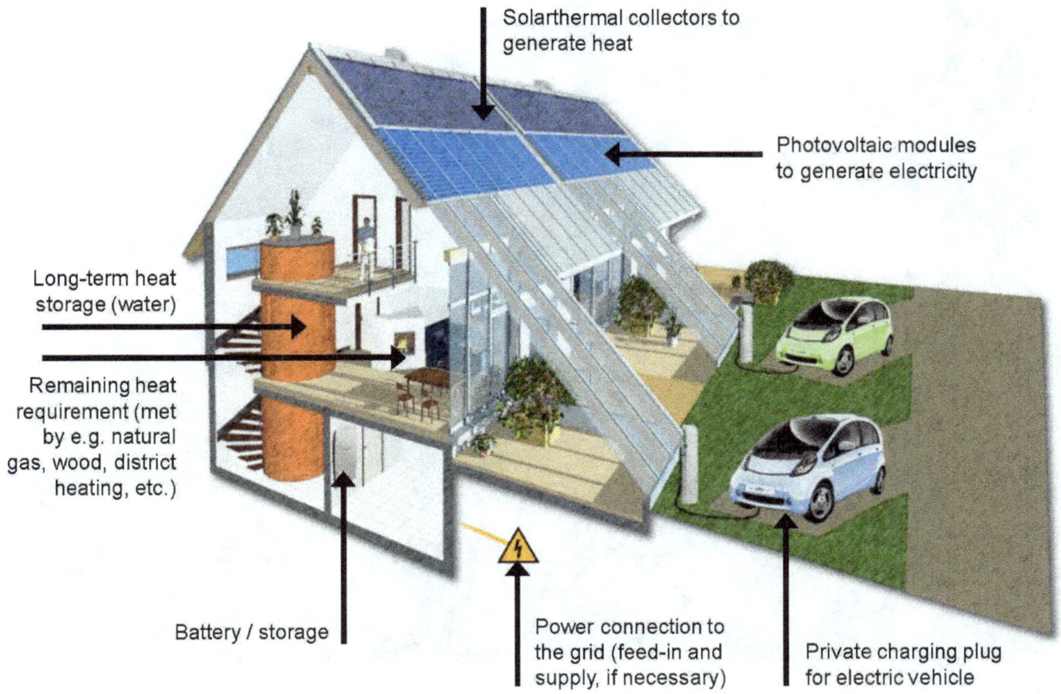

Figure 23: Features of Timo Leukefeld's energy-efficient houses.
Source: Leukefeld (2018).

energy services – including individual passenger transport – are significantly lower than the energy costs of 'passive houses', which achieve the bulk of their energy savings from passive sources, such as the heat produced by humans or technical devices.

The storage battery of my model house becomes part of a Smart Energy Grid with the objective to positively contribute to network stability and earn money from grid-balancing services. Instead of sending excess renewable power supply to neighbouring countries for free, or even paying money to get rid of the electricity generated by intermittent renewables – often accompanied by negative prices on the German wholesale market – the available electricity can be locally used inside the house for heating up the water in the reservoir. I favour lead batteries over lithium-ion batteries because they are still more economical. One can purchase lead batteries for €250 per kilowatt hour (kWh) of storage capacity. By contrast, in our current projects we purchase lithium-ion batteries at a price, depending on the size of the battery, of around €700 to €800 per kWh, including value-added tax and installation.

Of course, when lithium-ion batteries will become cheaper in the near future, I would favour that technology, because those batteries have a longer life-time and better performance, and generally are better suited for electric mobility.

Figure 24: Energy-autonomous, single-family house in Freiberg.
Source: Leukefeld (2018).

According to my assessment, the battery of the associated electric vehicle is too small to participate in the balancing market. By contrast, most of the electric vehicles that are currently available have batteries with a capacity of 12 kWh, but to effectively participate in the market they should contain a storage capacity of at least 50 kWh.

My team has expanded its planning approach from a single-family home to a residential building with seven flats. In this building, a battery of 54 kWh is installed, which may also be used for grid-balancing services.

Our energy-autonomous house has gained attention from media: Together with one of Germany's major publicly financed TV channels, I have prepared a 30 minute report about our concept, and many German newspapers and magazines have visited or invited me and publicised our insights.

The energy-autonomous house is an element of decentralised energy supply that has great potential but also faces major obstacles in its implementation. So far, the experience with zero-energy or plus-energy houses – that means, houses that produce more energy than they consume – has yielded mixed results. For example, the German Federal Ministry for Construction, Traffic and Urban Development (BMVBS) has screened and assessed 35 housing projects that are supposed to have an overall positive (both primary and final) energy balance.

According to the assessment, most of the houses have failed to meet expectations, though. The most frequent flaw is with heat pumps that are operated in a sub-optimal way (see also Schwoof 2013). For example, one observed house had a heat circulation system that was designed to provide the same temperature level throughout the entire house. However, the family that inhabited the house wanted to have cool bedrooms and a warm living room, and they changed the heating system accordingly, which led to a substantially higher consumption of electricity than expected.

Other flaws are related to the technologies that are deployed. BMVBS remarks that most heat pumps based on air temperatures performed worse than heat pumps based on soil temperatures or ground water.

My second idea is to favour quality over quantity. In my opinion, the German energy transformation (*Energiewende*) has been too focussed on quantity: producing a maximum amount of green energy, irrespective of the timing and location of feeding it into the grid. When the quantity of undifferentiated feed-in reaches a critical threshold, policies must switch to a more qualitative approach that takes the 'where' and 'when' into account. Annual energy balances that show a renewable energy surplus are often flawed – similar to so-called plus-energy houses, which may have a net zero-energy balance because of a surplus of energy production in the summer – but they would still require a lot of energy from the grid during the cold and dark winter months.

Heat pumps for residential housing are an expression and symptom of the 'quantitative' approach. Their promotion is a flawed policy incentive because they tap electricity from the grid just during the months when there is very little solar power available, thus reinforcing the need for conventional backup capacity. According to my calculations, heat pumps consume four-fifths of their annual electricity total during the winter months, whereas PV panels produce four-fifths of their electricity total during the summer months. We call this phenomenon 'seasonal illusion'. A heat pump cannot be operated using a typical residential PV installation during the scarce sunshine hours that characterise northern European winters. It may lead to positive results in the overall annual energy balance, but the anti-cyclicality contradicts the objective of achieving *qualitative* autonomy, which means the house is energy-autonomous during almost the entire year.

3.4.2 Business model

By April 2016, three players had already implemented our concept.

The first business model was developed by a group of housing associations, construction companies, and private real-estate investors. Their business model is based on a 10 year contract between the company that builds and owns the house and the people who rent it. They agree on a guaranteed rent over the entire time horizon, which includes living in the flat, heat and warm water supply, electricity, and – in the near future – electric mobility.

According to my experience, housing associations have to calculate with around €380 per square meter. For a multi-family dwelling with seven residential units, there is likely to be an additional cost of around 20 per cent, compared to a conventional building. Depending on the location, monthly rent amounts to €10 to €15 per sq. m, including the leasing of an electric vehicle.

The persons who rent the flat or the house face approximately the same costs as if they lived in a conventional house and had to purchase their electricity, heat, as well as the petrol at the fuelling station for their cars. However, they enjoy a much higher quality of living. In addition, the houses have barrier-free access. They also contain a lift for assisted living requirements and many more amenities. They also know with certainty that their all-inclusive rent will remain stable over the first 10 years.

The second business model is for energy utilities. They become a modern contractor and plan and operate all energy- and transport-related services for the multi-family building. I believe that our generation has entered the 'flat rate generation', which favours fixed payments instead of separate bills. Hence, people who rent the flat, and pay a flat rate.

This all-inclusive package substitutes for separate payments for heating requirements, electricity consumption, the leasing of a car, and the fuel at the fuelling station. Under very special circumstances – say, if collectors of amphibians have a set of energy-intensive terrariums and their consumption exceeds a certain pre-defined threshold of 'normal' usage of energy – they have to pay for the additional consumption at normal market prices, which is a payment scheme similar to contracts for internet roaming on a mobile phone.

Since the utility owns all energy-related installations, it can use them for various additional and revenue-generating functions, hence combining the management of assets and data as a service model. For example, all excess electricity and heat produced during the hot and sunny summer months can be fed into the electricity grid or sold to neighbours via a local district heating system. On days when the tenants do not need the electric vehicle that they lease, it can be rented out to other parties.

All of our projects operate with some involvement of an electric utility. In a number of projects, they just act as a service provider, in others they are the main contracting party. Utilities are changing their business model 'from selling commodities to services'.

The third business model is being pursued by banks. Two levels of involvement are attractive to them. First, they offer single-family homes. The typical target customers are people who want to have retirement plans that bring larger financial benefits than through conventional savings. In Germany, these savings are heavily taxed, so in the end, revenues after one's retirement are strikingly low. Instead, if one invests a similar amount of income into energy savings via a turnkey autarky package, these investments are not taxed when people get older. In sum, overall monetary benefits are twice as high as a conventional

Figure 25: Planned multi-family building in Rostock.
Source: Leukefeld (2018).

retirement plan, and people use their house as a guarantee for maintaining their standard of living when they stop working.

The bank typically steps in when people intend to build a house. The bank then offers to extend the credit in order to finance the measures to enhance the energy autonomy of the house. Via these investments, the future home-owner saves money – in the case of the model house that was monitored for its performance over three consecutive years, savings of around €3500 per year for the integrated energy and mobility solution were recorded. This is a sizeable amount of money, taking continuous energy price increases into account, and extrapolating annual tax-free savings from the moment the owners move into the house until they pass away – in Germany at the age of 81 years, statistically speaking.

In April 2016, a regional bank in Thuringia started one project of this type. The bank will promote the house over the course of three years. Its intention is to demonstrate how one can achieve high-quality living after retirement.

Banks also get involved in the business model based on multi-family homes. The bank then acts both as a financier and as the landlord of these residential units. It offers a 10 year lump-sum contract for potential tenants. Simultaneously, it offers a 10 year investment opportunity to potential donors with a very attractive and secure interest rate.

This investment opportunity also has a strong ethical component because investors know exactly where their money is going, they know the location of the building, and they can identify with their investments. One project of that type has also been launched by a regional bank. Its director reports that demand from investors is indeed very high.

In my concept, I have abandoned principles that have been common practice in my industry. Most notably, I do not want to rely on government subsidies for the financing of my projects. If credits or grants from, say, the German Development Bank (KfW) are available for specific projects, they are of course integrated into the financing scheme.

3.4.3 Scaling

In April 2014, my consulting practice counted fifteen projects, of which eleven had been initiated by housing associations, two by banks, and two by energy utilities. The six-unit block offered by a bank received 50 applications from potential tenants. Instead of investing €1 million in a conventional housing block, the bank invests €1.2 million. My team analysed the demographics of the people who sent enquiries about the housing projects and found that a large number of them were elderly and retired people.

The typical process of acquiring new customers involves an initial presentation at the partner institution, for example a bank, followed by workshops with the executive board and different business units, such as credit advisors, technical staff, and so on.

My consulting practice then enters the concrete planning phase for energy-related installations. In addition, we support the institution in its marketing efforts and in the promotion of the project by targeting customer segments which we would consider interested in the project.

It is very likely that the business model might be further refined and optimised, for example by extending the payback horizon and thereby reducing the rent and leasing costs, after a number of residential units have been in operation for a couple of years.

Another future option is to integrate commercial buildings and industrial sites into my practice's portfolio. As long as the customer is not in an energy-intensive industry, such as an aluminium smelter, our concepts would work in the same way as for residential buildings. Supermarkets with standardised buildings and commercial retailers would be potential candidates. With an increasing number of projects and publicity in the media, it might only be a matter of time until a player in that field approaches us. Working with large corporate clients might pose risks for a small consulting practice like ours, though. Their corporate culture might be non-compatible with our ideas of cooperation. If the power relation is too asymmetric, for example, the client might exert pressure and force us to reduce the price of our services.

The market for retrofitting existing building stock is much larger than the market for greenfield constructions. Around a quarter of residential homes in Germany would be suitable for renovation in order to become energy-autonomous, according to our estimates, because these houses have large, uninterrupted rooftops facing south that could be used for solar heating and PV panels. However, the retrofitting of these houses faces two main barriers. First, investors would like to implement standardised solutions, but for a renovation, one has to take the diversity of existing buildings into account. Each renovation has to be adapted to the individual requirements of the object. Adequate insulation and modifications to the layout of the flat are basic requirements for a properly executed retrofit, which drive costs up.

The second barrier is related to rights and privileges that tenants enjoy: A renovation may increase rental costs by 100 per cent, but tenants are well-protected under German law. They would have to leave the flat and be replaced by new tenants willing to pay the required rent. However, especially if old tenants have inhabited the flat for a sufficiently long period of time, it is almost impossible to force them to leave under German law. They may successfully use the legal framework to oppose that increase in rents or the threat of getting kicked out.

With a renovation rate below 1 per cent of the existing building stock in Germany, apparently all relevant actors are failing to establish a standard solution for how to increase the rate of renovations.

According to my assessment, both commercial buildings as well as the retrofitting of existing housing stock will become relevant and attractive once the market segment with the best framework conditions – in particular, new residential units and new single-family homes – has been tapped. My consulting practice will have climbed further along the learning curve through the experiences gained from ongoing projects. For instance, one of the learnings that has already taken place concerns the configuration of the water tank, which was modified in its design in order to be more easily integrated into multi-family homes. These improvements in design could also be applied later in the configuration of commercial buildings, such as supermarkets.

3.4.4 Market outlook and competitive environment

Competition in the field of energy-autonomous houses is limited. A number of architectural practices and construction companies offer turnkey solutions. Pilot projects are often led by universities or start-ups based on academic research. Emphasis may not be exclusively on energy autonomy. For example, Graft Architects offers a plus-energy house that follows the cradle-to-cradle principle, which means that all materials can be fully recycled. It is built with 'healthy', organic materials, such as wood and clay.[52]

[52] Source: Personal interview and subsequent email exchange between the authors and Lars Krückeberg (22/4/2016).

From my experience, I believe that energy-autonomous houses can be commercially successful. If we are able to acquire even more customers in Germany, it is likely that competitors in Germany and in other countries will follow suit. Reinforced by regulation, energy-autonomous houses will become the standard rather than the exception in the newly built urban and rural environment. However, these new structures constitute typically only around 1 per cent of total housing stock in industrialised countries. The major challenge for decentralised energy will be the renovation of the 99 per cent of *existing* housing stock.

3.4.5 Interviewee biography

Prof. Timo Leukefeld studied energy engineering, worked in academia, and then founded his own company, which planned and installed solar heating systems across Germany. In 2011, he switched to consulting activities, leaving the actual implementation to other players. His current work is based on three pillars: (1) applied research and development, including the position at the TU Bergakademie Freiberg through which he holds lectures about energy-autonomous buildings, (2) the concrete planning of either new or refurbished energy-autonomous residential dwellings or commercial buildings, as well as (3) offering keynote speeches across Europe. He also serves as a strategic advisor to the federal German government and to commercial actors such as banks and housing agencies.

3.4.6 References

Schoof, J. (2013), *Von der Schwierigkeit, ein Plusenergiehaus zu bauen, Detail, 14 August 2013*. Retrieved February 11, 2019, from https://www.detail.de/artikel/von-der-schwierigkeit-ein-plusenergiehaus-zu-bauen-10920/

3.5 Entelios: Demand Response – a decentralised approach to complement intermittent renewable energies

Interview with Oliver Stahl, founder and former CEO of Entelios, on February 11, 2016

Entelios is one of Europe's leading energy management solution providers for decentralised energy resources in the industrial, commercial, and institutional sectors. Entelios provides Demand Response services for large commercial and industrial energy consumers. Additional clients are European energy companies in need of value-adding energy services for their B2B customers, or interested in enhancing grid and supply stability through Demand Response. As a partner, Entelios enables those companies to set up their own Demand Response programmes. Based on its leading solution suite and operational capabilities, Entelios offers automated industrial 'Demand Response As a Service'. As one of the first Demand Response aggregators and white-label Demand Response solution providers in Europe, Entelios has been building operational expertise in Demand Response since 2010 (Entelios 2018).

3.5.1 Technology and business model

Entelios was the first European-focussed Demand Response service provider that offered fully automated Demand Response for distributed energy resources (loads, generation units, and storage) in industrial, commercial, and institutional settings.

Technology

Together with two colleagues, I founded Entelios in 2010. I am a serial entrepreneur who worked for quite a number of years in strategy consulting. I studied electrical engineering as well as business administration, among other subjects, at the University of Mannheim. I also worked in robotics and automation and did my MBA at the Sloan School of Management at the Massachusetts Institute of Technology (MIT). While I was at MIT, in 2008, I was in search of a new business model that would fascinate me. I was given the opportunity to co-design and help organise a course called 'Energy Ventures' at MIT. For this course, we invited numerous CEOs from the energy sector and analysed where new business opportunities could emerge. Among the CEOs we invited was Tim Healy, the co-founder of the US company EnerNOC. EnerNOC has existed since 2001 and went public in 2007. During his lecture, Healy introduced us to the concept of Demand Response. In his business model, he pools and aggregates the flexibility of industry players to increase or reduce their electricity demand – and also their supply, in case those industrial players have distributed energy resources (many have, not only for backup purposes). By selling these modifications in demand to the grid operator, EnerNOC financially compensates the companies that participate in the Demand Response scheme.

Business model

After my return to Europe in winter 2008/09, I had numerous discussions about the idea of setting up a similar business with established players in the energy industry. When I intend to launch a new business, I do not start it by developing new software. Rather, I sign a contract with a larger industry player fairly early in the process. I present the idea and key milestones, which may be discussed with industry partners, but *de facto* I expect that the financial foundation of the new venture will be solid by then. For me, having customers early on is essential! If they are willing to pay, that is close to a 'proof of concept'. With this approach, a start-up does not depend just on classic bootstrapping, in particular financial support from business angels and venture capital funds.

Stadtwerke München, the municipal utility of Munich, decided fairly early on that they wanted to head in the same direction as proposed in my idea. They were already operating a 'virtual power plant' (VPP), in their case a certain number of small power-generation units (*Netzersatzanlagen*), which they combined into a larger, virtual generation plant. For their VPP, they used generation units that they already had, colloquially speaking, in their backyards. We presented the idea to them, in particular the idea of targeting industry and commercial players, and convinced them. They then became our first industry partner. Of course, we later received support from venture capital funds, but most important for me was that we were able to convince a large industry player of our business idea, and Stadtwerke München had already signed a contract before we even approached venture capitalists.

We then talked to the transmission grid operators and convinced them that our ideas have a solid and beneficial basis. It certainly helped us that the business model was already established in the United States and that Stadtwerke München supported us. For one year, we developed our software in collaboration with Stadtwerke München as our partner. We did the first tests together with Stadtwerke München and the transmission grid operators. We also cooperated with the Forschungsstelle für Energiewirtschaft in Munich, one of the oldest German institutes doing research on energy efficiency. They helped us to comply with all technical standards and regulations. After one year, we received the necessary certificates, accreditations, and the approval that all of our operations were in compliance with the so-called transmission code, which sets the standard for the German market.

We sell our Demand Response product on the market for balancing energy. This market had been traditionally dominated by large generating units operated by established energy players. Then we came along and intended to do the same, but with hundreds of small entities (distributed energy assets) that we combined. They are aggregated with customised software – up to the precision of one minute – and are in compliance with the requirements of the transmission grid operators.

This took us two years. During that process, we received the first Series A of venture capital. We had a lead investor from the Netherlands, plus the High-Tech Gründer Fonds – the largest early-stage investment fund in Germany – and a business angel from the Netherlands. This combination gave us a financial boost, but it took two additional years to become operational because we had to overcome numerous hurdles, most of them related to the complexity of the regulatory context.

The key question for business models in the so-called Smart Grid space always is: Who pays for the products or services provided? Demand Response encompasses three types of decentralised energy assets, which provide flexibility for the electricity system – either generating units, energy-consuming devices such as lighting, or energy storage, which can be conventional galvanic elements, that is, batteries, but also other types of storage. For example, an aluminium producer may have the flexibility to control its production of aluminium blocks, which are, to some extent, stored energy.

Many contracts are needed to deal with the transactions that occur in Demand Response. Ultimately, we take flexibility in energy consumption or production from one player in the market and give it to another one. In this process, we have to negotiate contracts with our industry partners, of course, but also with operators of the high-voltage transmission grid and the lower-voltage distribution grids, as well as the administrative unit in charge of the regional balancing processes (*Bilanzkreisverantwortlichen*). With each of these partners, four to five contracts have to be signed individually, which adds a certain degree of complexity to the process.[53]

Paulaner Brauerei, a traditional Munich brewery that belongs to the Schörghuber Corporate Group, was already active in corporate social responsibility and pursued a strategy of environmental awareness and protection. Hence, they also agreed to become our first major corporate client. We took their assets and connected them with our software. The contracts that we formulated with them later served as a reference for other clients. We took one company from an industry sector, such as Paulaner, and with that reference it was easier to approach their industry peers.

---

[53] Since the interview took place, new processes have been established that have simplified and improved the legal procedures.

Not all industry sectors are equally attractive. The most difficult case was the chemicals industry. Processes there are already highly automated. Tiny changes in temperature can completely distort a chemical process. We went into an intense dialogue with industry players to find flexibilities. For example, a heating system based on oil or natural gas can be transformed into a bivalent process, with electricity as an alternative energy source. Excess electricity can then be absorbed by the system at short notice by switching the energy source from oil to electricity or gas to electricity. The companies became aware that just through minor modifications of a heating system they could have annual revenues of around €20,000 up to €80,000 per MW. It then took another three months to modify the facilities. In some cases, we also took over the costs to accelerate the process. With these incentives, we eventually also convinced players in the chemicals industry.

After six to seven years of actively promoting Demand Response, the industry eventually started to change its mindset. Even some of Germany's very large industrial electricity consumers are now interested in flexibility options and new business models. They want to actively engage in the energy transformation and earn money with flexibility services so that they can remain competitive with their core products on the global market.

Money to build a company is easily made available as long as there is a meaningful business model. It is not easy in the electricity sector because it is such a highly regulated market. There are many established players. For example, municipal utilities are currently in the process of redefining their strategies. Large utilities used to earn their money by selling electricity as a commodity. This has largely disappeared. Final customers are moving towards energy autarky and one day may no longer want a connection to the grid. Medium-sized companies are installing decentralised generation units in their backyards, such as small combined heat and power generation (CHP) units.

We moved into two strategic directions: On the one hand, as an independent aggregator, which is a completely new role in the market; on the other hand, we are a service provider with our software for larger utilities. We call it 'Demand Response As a Service'. In the first line of business, we directly approach industrial customers and aggregate their flexibility potential. In the second, we have software and a back end that we call the Network Operation Center (NOC). Traditionally, large utilities are not used to thinking in terms of megabytes. For them, it was always 'make', sometimes 'buy' – but hardly ever 'lease'. Software as a service was not part of their mindset.

3.5.2 *Scaling and cooperations*

We have identified the top five industry sectors in Germany, including aluminium, silicon, chemicals industry, paper production, and breweries. We are looking for industries with a high degree of automation and annual electricity consumption of more than 10 to 20 GWh. The top management CXO suite has to be aligned behind the idea because support of the procurement, where electricity purchases traditionally have been located in the organisation, is not enough. We have to convince the CEO, the operating managers of the facilities, and the CFO.

All facilities that have decentralised generation units, such as industry parks, are potential clients of ours, too. Water supply is also interesting because of the pump systems. There are hundreds of pumps, similar to gas pipelines. One has to become creative.

Scaling

We established one reference company in each relevant industry sector and then convinced their peers as well (reference selling).

We could build upon, to some extent, the established role model in the United States, but they have a different approach. Across the Atlantic, the main driver used to be the prevention of blackouts and brownouts, whereas in the German context, the driver is the integration of renewable energies and the mitigation of the volatility of those renewable energy sources.

In future energy systems, energy procurement will still have its role. Imagine a company that outsources its entire energy management. Energy Intelligent Software (EIS) is the key, and we can provide all services regarding procurement, efficiency, Demand Response, and billing end-to-end. It is not a core competence of industry players to deal with energy procurement, although many became experts in that as well.

In the energy-intensive industry, for example paper production, it makes sense to have an energy procurement unit. They sometimes even have dedicated trading floors. But for many other industries, it does not make any sense to entertain such a trading unit; they could easily outsource it. In the United States, we approach clients and suggest taking over these functions, or we provide adequate experts from our side who are either dedicated to one player or a pool of players.

Multinational companies receive hundreds of energy bills. We offer a solution that manages all of these contracts and matches the actual consumption with the billing processes. We currently have roughly 1 to 1.5 billion data entries in our IT centre every month. That is a Big Data business model. We see a company's options for flexibility on a granular level that companies would typically not be able to see themselves. We can detect patterns in the energy consumption of particular energy assets, which gives us the transparency to predict how the device will behave at a specific hour of a specific day. When thousands of those pieces of information are combined into an aggregated flexibility package, the package can be sold. But we can also do predictive maintenance because probabilities in forecasting are high. If a device behaves in a strange way, that may be a hint that there is a planned or forced maintenance event, or even a shutdown or malfunction. One could add many business models on top of Demand Response.

Currently, our most important challenge is to grow in the area of software and IT. On a global scale, there is no real software that provides a full-fledged end-to-end solution in energy management that is similar to solutions for customer relations management or sales management. In this field, we want to position ourselves in the large enterprise segment. It is our objective to organically grow in this field.

Cooperations

Energy utilities are still sufficiently well-endowed to acquire companies and their proprietary technologies. Acquisition activities occur in cycles. Investors behave like lemmings: If a Demand Response provider is acquired, many other players in the same industry try to acquire similar companies. Our competitor in France was acquired by Schneider Electric; another competitor from Austria and Slovenia was bought by a Japanese company; a Scottish company was acquired by a Swiss player. Then large industry players such as Siemens and General Electric became aware and also wanted to become active in this field, in particular to complement their activities in the Smart Grid. We were approached by the industry. We had one offer that we considered as not being sufficient, so we employed a boutique M&A consultancy to prepare our company for a potential takeover and invited a larger circle of potential acquirers. We realised that there are fundamental differences between the cultures of different companies. We had to ask ourselves whether we just wanted to become a business unit in a large multinational company with a workforce of 100,000 employees. It might then not be clear whether we would still be an independent business unit in one or two years.

By contrast, EnerNOC has Demand Response as their core business model, so it was a win-win. My management board gave me the freedom to decide who to partner with. Obviously, we also had many discussions about who was the 'right' partner. EnerNOC had tried to enter the Demand Response market in mainland Europe, maybe because they were too fast and somehow too 'American' in approaching, for example, traditional German utilities. However, they successfully entered the markets in the United Kingdom, Australia, New Zealand, and even South Korea. In addition, they had a very similar corporate culture and mindset, and we knew them a bit already. The story was just right.

Since the acquisition, the software aspects have become more important. When we were on our own, we did not have the entire software suite, which enables us to provide end-to-end energy management for our clients. We had software that was, in my opinion, world-class in terms of Demand Response. In other regions of the world, Demand Response is much slower and has longer lead times before it kicks in. Our software was much faster, which is why we were also attractive for EnerNOC, but we did not have all the other elements. If we sell a Demand Response product to a company, we are asked very quickly whether we can also help in other aspects of energy management. One of our competitors that was acquired by Schneider Electric one year before us already had the whole software suite after the acquisition. It became a difficult situation thereafter from the perspective of competitiveness.

We also had a partnership with E.ON. But when a large player like E.ON joins forces with a start-up like ourselves, it is somehow also a proof of concept that our business model and our software solution are leading edge. Depending on the pre-defined milestones to be achieved, we also received financial support from them. The collaboration was handled mainly by E.ON's sales unit, but sales are driven by the personal objectives of each key account manager to sell more power and gas, and Demand Response is just a small side product that only makes a marginal contribution to their target agreements. It takes time to establish that process and to formulate the templates for the contracts – often more time than a start-up actually has. We had to see success at a faster pace. It helped us to learn how to make our product more complete.

Now we have a sales unit for the utilities, and we have key account managers for individual industries, such as the paper industry. I am looking for sales managers who have access to CXO suites and a track record in high-level sales, someone with expertise in electric devices and a successful sales profile.

Energy utilities are scared of what could happen if Entelios ceases to exist. For instance, they would have liked to acquire Entelios shares, but I wanted us to be able to provide services also to their competitors in the future. Nonetheless, we had to cooperate closely with larger utilities. In one instance, we were very disappointed when a player then tried to copy our product. But we had long-term cooperations with faithful players such as Dong from Denmark, Verbund from Austria, BKW from Switzerland, and the Stadtwerke München. They realised that they would not be able to manage the energy transformation by owning all the intellectual property and software, and they made efforts to find partners with whom they could jointly develop new products and services. Together, we have to build an ecosystem and a peer-to-peer landscape.

It is unclear where Entelios fits into the organisational structure of a traditional utility: The sales units have contact with the end-customers, but there are also the generation unit, the innovation department, and the trading unit that would potentially fit. Utilities did not know where to position us because we provide a product for the market for balancing energy and managing the consumption as well as the production of energy assets, hence we should belong to the generation unit. But we also do trading, and we directly approach the final customers in the industry and aggregate their flexibilities in energy consumption and production.

Fairly early in the process it became clear that we should be linked to the sales department if we wanted to be successful. For the sales personnel of a utility, who have been selling energy over decades, it is difficult to sell the opposite – a reduction in energy, which is a 'totally different animal'. In the beginning, only very few sales managers were willing and demonstrated the 'mental flexibility' to focus on the new concept of Demand Response.

Interestingly, large utilities launched spinoffs, new ventures that were 100 per cent owned, of course, by the utilities, with some internal but also external staff that were targeted directly for developing new business models for the energy transformation. I believe this is the right way. Those who have always worked in the old energy world have had a harder time imagining that their old world may one day completely disappear. Few understood that customers had to be approached in a completely new way.

For example, an industrial client may want to launch a new product. All the equipment that concerns building efficiency, energy provision, and all the technical details may no longer belong to its core competencies and strategy. When we started, municipal utilities did not have the idea of building and/or providing services, for example for a combined heat and power (CHP) unit and leasing it to their industrial clients. In contrast, where would a municipal utility earn its money with an increasing number of residential customers who do not need a grid connection, because they have become energy self-sufficient? Utilities need to think of new business models. For example, municipal utilities could build new housing units and equip them with all energy-related technology, including PV panels, heat pumps, and all other elements of building efficiency. That utility may then no longer charge for electricity or gas, but rather a fixed, all-inclusive rent for the individual flats or housing units. A frontrunner for those novel concepts is the Freiberg Institut (www.freiberg-institut.de/), under the leadership of Dr Timo Leukefeld.

3.5.3 Market outlook and competitive environment

Imagine walking into an industry player's facilities and telling the management: From now on, I want to steer and control your assets. Then all sorts of resistance emerges. For example, the people in charge claim that their facilities are already highly energy-efficient. Companies also raised questions about the reliability of our operations, what would happen if we damaged their assets, and how much we really earned with our business model.

It is a paradigm shift. For more than 100 years, supply just matched demand, but now supply has become increasingly volatile with the deployment of renewable energies. When the wind blows at 4 am, residential customers are still asleep and industrial processes have not yet commenced. Today, the wind rotors have to be shut down, but in the future, we will explore opportunities to temporarily store the power. For example, at 4 am, all air pressure compressors (*Luftdruckkompressoren*) in Germany could be filled with air. Compressed air just acts as an intelligent storage device, similar to a block of aluminium or silicon.

The emerging business of aggregators managing flexibility and the underlying energy assets are typically active in biomass with small, decentralised generation units, but they are operating in a completely different business model. We have observed large revenues but small profit margins. There are indeed already three larger aggregators that are active in load management and storage. An upsell is difficult for these assets. Selling additional services to a rural community with biomass assets is difficult. We decided that we would not head into this direction, but instead focus on industry and commercial businesses.

In our nascent industry, competition is very healthy because it gives the impression to potential clients that there is an attractive market, and in the end the overall size of the market increases.

In the United States, establishing a new business model takes less time than in Europe. When brownouts were looming, people literally called large consumers via telephone to shut down parts of their assets. They were much more pragmatic than here in Germany. Demand Response in the United States works via email, text messaging, and call centres, as opposed to our highly automated system in Europe. Second, politicians very quickly embraced the concept, and the US Congress ratified a Demand Response action plan, which led to a modification of the sector's regulations and legislation. With this move, the role of the market aggregator became legally accepted. Third, large associations were founded, and lobbying activities were more pronounced.

In Germany, we have to be grateful to some extent to the Green Party for promoting renewable energies. But we could also blame them for not having thought about the wider system, for example adequate storage solutions, transmission lines running north to south, and the reinforcement of the grids. It was a mono-dimensional approach, and now we have a transformation (Energiewende) with huge flaws and ineffective markets. Above all, the self-imposed CO2 reduction targets cannot be achieved.

When we talk with associations about municipal utilities, they recognise the value of our business model, but they do not want to actively push it because they want to allow their subsidiaries to develop their own technologies and become market-ready.

In Brussels, we co-founded the Smart Energy Demand Coalition (SEDC, which is now Smart Energy Europe, www.smarten.eu), together with other innovative energy players and stakeholders. The former EU commissioner for energy, Günter Oettinger, quickly grasped the benefits of Demand Response. Not long thereafter, Demand Response was part of the energy-efficiency guidelines of the European Commission. In Germany, back then in 2014/2015, this had not yet been implemented into national regulations. This was probably because Germany had so many other questions to deal with on a national level, but it is clear that a further expansion of renewable energies – combined with a shutdown of all nuclear assets by 2022 and all coal plants by 2030 – without Demand Response will face challenges in terms of a stable supply situation. If no wind or sun are available, we have to have a backup option for the domestic baseload of around 80 GW. Alternatively, we would have to install adequate storage capacity – but where should this come from?

Renewable energies cannot yet completely substitute for traditional power plants. In particular, we are lacking the transmission lines to transport electricity from the north to the south, but more importantly, we do not have sufficient storage capacity. One could imagine the gas grid with power-to-gas, or Demand Response, or stationary grid-scale batteries, or even batteries for residential households, as it is currently evolving in the United States. We need all these options, but we should have started much earlier, developed a comprehensive plan, thought about these options, and pursued R&D activities in these fields. We should have been focussing on these topics since the late 1990s.

In the United States, large electric utilities claimed that Demand Response is a retail-oriented business model, but it operates with wholesale prices. The Federal Energy Regulatory Commission (FERC) allowed Demand Response to participate in wholesale markets, but some states claimed that this was beyond the FERC's authority. But the US Supreme Court decided in favour of the FERC's order. In January 2016, the Supreme Court of the United States announced that it sided with EnerNOC and the FERC in the case of *EPSA v FERC*. This gave EnerNOC a boost of confidence, which was reflected in its stock prices: It fell to US$4 and then rose again by 70 per cent on the day of the Court decision. If the Court had decided against us, a large portion of our portfolio would not have been able to participate in the capacity markets, and a large part of our cash flow would have been at risk.

3.5.4 Interviewee biography

Oliver Stahl was founder and CEO of Entelios AG (www.entelios.com)*, and later, after the acquisition of EnerNOC, Managing Director Europe for EnerNOC. He still serves as Senior Executive Advisor to the company. Since April 2017, he has been CEO and co-founder of Robotise (www.robotise.eu), a start-up whose mission is to introduce service robots into people's everyday lives.*

Oliver is a serial entrepreneur and was also a management consultant in a global consulting firm for several years. He is co-founder and a board member at SEDC, Smart Energy Europe (www.smarten.eu)*, a European industry association with more than 50 member companies with a focus on Demand Response, demand-side management, and energy efficiency.*

Oliver started his career as an electrical engineer in the area of industrial automation and robotics. He studied business administration and educational science at the University of Mannheim, then at Loyola University and Northwestern University in Chicago. In 2008 he became a Sloan Fellow and received an MBA degree from MIT Sloan School of Management.

3.5.5 References

Entelios. (2018), *Consumption follows generation, demand follows supply.* Retrieved February 10, 2018, from http://entelios.com/entelios/

3.6 SOLshare: decentralised energy supply – complementary or antagonistic to rural electrification

Interview with Sebastian Groh, Co-founder of SOLshare, on January 9, 2017

SOLshare is a Bangladeshi Ltd. ICT company founded in 2015 that has created a revolutionary new approach to bring affordable solar electricity to everyone in Bangladesh and beyond through its peer-to-peer solar energy trading platforms. SOLshare is at the brink of creating the next generation of electricity grids. The social enterprise pioneers a micro-energy transition model 3.0 interconnecting solar home systems, monetising (excess) solar energy along the value chain in real time with mobile money, and empowering communities to earn a direct income from the sun (SOLshare 2018).

3.6.1 Technology and business model

In 2009 I started working at MicroEnergy International, a Berlin-based consultancy specialised in decentralised energy services. We first looked at sharing electricity from an academic perspective. I did my dissertation on the topic of innovation and energy service supply along certain development processes as an economist, and a colleague of mine made contributions from an engineering point of view. We basically found two things. First, when a solar home system is designed, there is a trade-off between the rainy season and dry season. This means that either the battery has sufficient capacity to capture all of the power generated by the panel (but then it will never be fully charged during the rainy season), or a smaller battery is provided, albeit with the disadvantage of it never reaching full autonomy during the rainy season. After two days of constant rain, the system will go offline. In Bangladesh, we would like to have at least three days of full autonomy. We had to settle on somewhere in the middle. That also implies that around 30 per cent of what the panel could produce over the year cannot be stored by the battery and goes unused, which is 600,000 kWh daily! That is the equivalent of driving a Tesla Model S 68 times around the globe every day. There is a massive potential available across almost five million solar home systems just in Bangladesh.

Our second observation was that there were already lots of people sharing their electricity, for example by going to a neighbour and asking them to charge a phone, or there was someone who did this on a commercial basis and took a certain fee for that. But we also saw people connecting their households with cables to their neighbours and paying per hour of light bulb use. They had to monitor this at all times, though. Translating these fees into kilowatt hours opened our eyes: from US$3.50 up to US$10 per kWh is what we economists call the energy-poverty penalty, and what we entrepreneurs call a massive market opportunity. Combining these two insights creates an opportunity in which everybody can win. People are willing to share and pay for power, and there is excess electricity available. Plus, there are still many households that cannot afford a solar home system, despite the micro-credit schemes. However, the sharing scheme can be designed to be inclusive enough to make it affordable for them as well, while turning other prosumers into local energy entrepreneurs.

In Bangladesh, there is an extremely high density of these solar home systems. Our idea was to connect these systems and, based on the connections, to increase overall capacity through synergy effects. At the same time, in order to achieve a certain level of diversification, we would tap into the unused 30 per cent potential. We started from this idea and called it 'swarm' electrification, in which the swarm is more powerful than the individual unit and we would have peer-to-peer communication. Each system speaks to the next system. If one system drops out, nothing happens – a fairly resilient model. In contrast to a conventional mini-grid, which has a determined system size, our model can also grow organically.

If we interconnect solar home systems and incrementally add more generation and storage sources, we will have a much better business case because our capital expenditure (CAPEX) is limited to the cabling from house to house, as well as to a bi-directional metre, which we call the SOLbox. It essentially works as a net metre. Whenever a household is a net consumer, its balance decrease – not in kilowatt hours, but in Taka, the Bangladeshi currency. Whenever a household is a net producer, the balance increases. The balance is directly mirrored on the mobile phone in a mobile money wallet. Then that money can be used in real time for payments of all sorts, not just energy. That is the biggest value proposition for our customers.

Technology

We develop the SOLbox in Bangladesh. Then we have a couple of suppliers in China from whom we order the printed PCBs (circuit boards). We produce the case in Dhaka and also do the assembling and testing there. Because of the evolution of the sector over the last 15 years, the local workforce is well-prepared.

The cabling and the WiFi tower should be covered by the membership fees. The partner organisations (POs) can get much better rates than us to purchase the cables due to their good relationships with the suppliers. We tell them when something is needed, then they buy it themselves and recover their costs, plus a margin via their subscription fees.

From the perspective of maintenance, our system is slightly more complicated than the average system. We have developed a training-of-the-trainer concept: We are training the regional managers of the POs to then train

Figure 26: SOLshare's field operations team providing technical support.
Source: SOLshare (2017).

Figure 27: Production facilities in Dhaka.
Source: SOLshare (2017).

their staff – also in terms of marketing. They recommend that people get a mobile wallet if they do not already have one in order to avoid having cash in the field.

Business model

The Bangladeshi solar home system sector really started in 2002/2003 with the launch of the Infrastructure Development Company Limited (IDCOL), the governing body of the programme, which is under the Bangladeshi Ministry of Finance. Since then, around US$750 million of development finance money has gone into the sector, all channelled through the IDCOL to bolster the ecosystem of solar home supply. Up until 2016, more than 4 million solar home systems had been installed.

We basically run a mixed business model that has B2B and B2C elements. Our first customers were the implementers of solar home systems. We piggyback on the existing infrastructure and cooperate with more than 50 partner organisations of the IDCOL, which provide both technology and financial credit to the customers in a one-hand model. Over the last 15 years, the World Bank and other institutions have given loans at very low interest rates to the IDCOL, which passes it on to the POs at a slightly higher interest rate. The POs then provide two- to three-year loan contracts to the end-users at an interest rate of 8 to 12 per cent.

Our SOLbox is an additional component of their solar home systems. We sell it to the POs, for example Grameen Shakti or Bright Green Energy Foundation, among others – the former being the biggest PO with more than 1.7 million systems currently installed. The PO then integrates it into their solar home systems. We also retrofit (upgrade) existing systems, but always in cooperation with a PO.

Our revenue stream is threefold. We first have a margin on the SOLbox, which we sell to the POs. Secondly, similar to an Uber business model, we take a fee for every transaction. When neighbour A transfers money to neighbour B, there is a spread, and this money goes into our mobile wallet. This is the B2C component of our business model. However, we share this fee with the POs – also to give them an incentive to keep the grid running and encourage further expansion.

The third revenue stream is a fee for managing the grid, which the POs transfer to us. They can decide by themselves how to charge the customers for this service. As we have WiFi towers in the villages, our SOLboxes communicate with the WiFi towers. We did not want to integrate a GSM chip into each box, which would have turned out to be too expensive. The WiFi tower is the single point of communication to us. With this technology, we can monitor at all times how much electricity is being generated as well as how much is being consumed and traded. With this information, we can determine which appliances are being used and which other appliances could be used with more

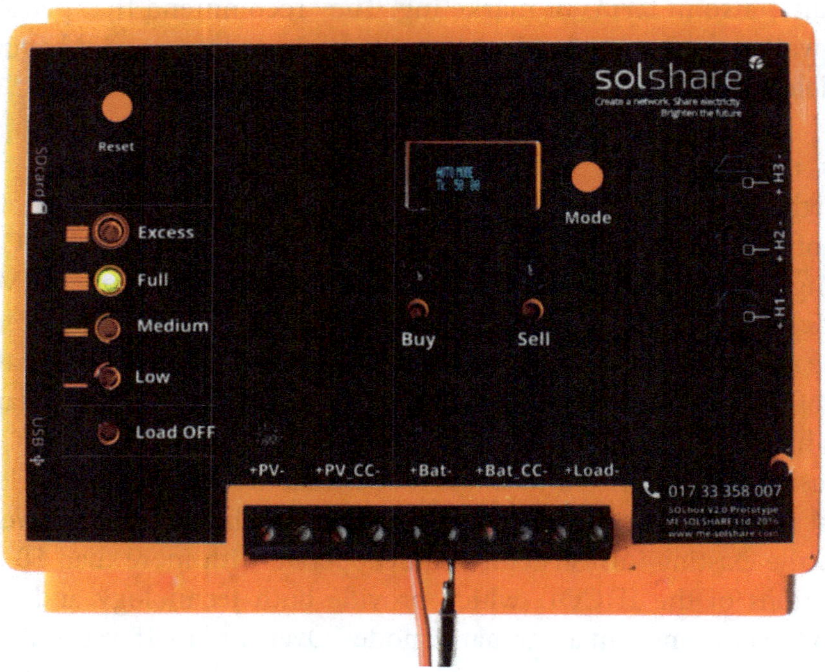

Figure 28: The SOLbox.
Source: SOLshare (2017).

generation capacity or more storage capacity. These diagnostics and analytics are sold as a service package to the partner organisations.

Typically in mini-grids, there is a connection or subscription fee, which is either one-off or recurrent. Together with the POs, we are still looking for the optimal business case. Still, a lot of people need to be convinced that they can earn money with their solar home systems. The general understanding is that it is purely a consumptive good, but suddenly we have turned it into a productive good. Second, it is a connectivity good that we are pitching – together with the POs – to the end-users.

If a residential customer buys a new system, it costs about US$360 over the period of a 36 month loan. That is US$10 per month (ignoring interest to make it simple). The SOLbox is sold for US$24 to the PO. The PO then puts a 20 to 30 per cent margin on top. We calculate a recommended end-user price of around US$30, which is less than a dollar per month of additional expenses. If customers sell 1 kWh for US$2 each month, they make a net profit from month one onwards. Obviously, it cannot be guaranteed that they will actually sell a kilowatt hour each month. Maybe when they want to sell, nobody will want to buy. To take that into account, we put a buffer system into the village – like a micro-utility, where the electricity is buffered. We then connected a water pump to the buffer system, which is a nice load because the water pumping can take place during the day and not so much in the evenings. The electricity can also be used for alternative devices such as corn shellers or rice cutters. Recently, we launched our first pitstop solar rickshaw charging stations, where rickshaw drivers can charge their vehicles during their lunch or tea breaks. In the evening this pays off, as they can ride their rickshaws an extra hour before they have to make their way to the closest national grid point.

In the traditional ABC mini-grid model (anchor, businesses, community) everything is designed for a specific anchor, usually an outside investment, which eventually takes out the return at some point. By contrast, in our system the grid is designed for the community. The community grows its own capacity to power anchors, and the returns stay within the community as well. So one could refer to it as a reverse ABC model. An average household in Bangladesh earns between €50 and €100 per month, the 'rich poor' in the villages, plus around €1.50 from selling electricity. But we are targeting micro-entrepreneurs: Knowing that one can earn money with our installations, local residents install extra panels and make additional profits. The price per panel is very cheap, meaning that one can make a lot more money. As a reference point, a kilo of rice costs about €0.50. With three to four kilograms of rice, one can get quite far.

We also have a second product. It is our retrofitting model from post-paid to pre-paid, which we call the SOLcontrol. There is a payment crisis in Bangladesh, which means the repayments on loans for the solar home systems have significantly plummeted; repayment rates have dropped from up to 95 per cent

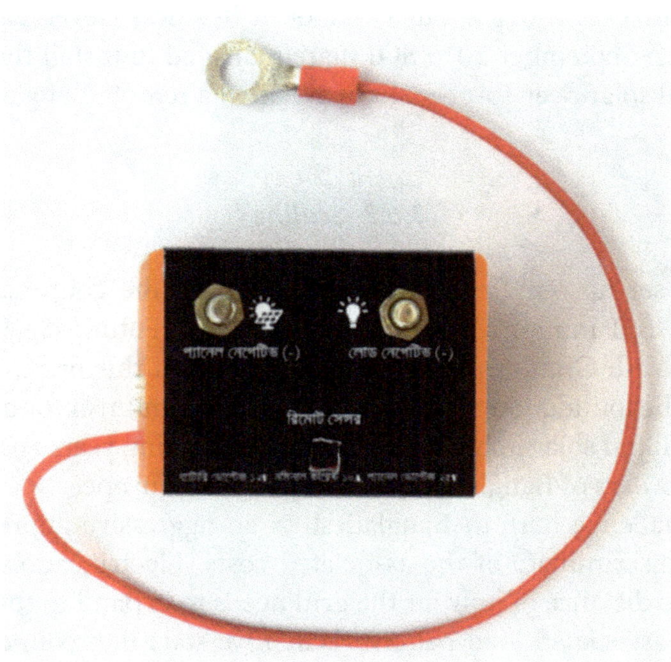

Figure 29: The SOLcontrol.
Source: SOLshare (2017).

down to 30 per cent. We are targeting this with two actions. In existing systems, we can easily retrofit them from post-paid to pre-paid to make them pay-as-you-go. We are thereby trying to help the POs to keep their good customers so that they do not – through a domino effect – become bad customers, and to help them turn some bad customers into good customers. Enforcement is also better if we can electronically lock the system rather than go into the village and uninstall the systems, which is in most cases not happening anyway.

The SOLcontrol is a good learning case for supplying 150 households in the first month, and 5000 in the following months. We will grow with this experience, because, together with the SOLbox, the business model will become more complex. We are diversifying our business model every day and really have to simplify and focus. We have a lot of options in terms of pricing, but we have not found the optimal model yet. Maybe we will scale up with a different pricing model.

3.6.2 History and organisation

SOLshare, located in Dhaka, Bangladesh, is a spinoff of MicroEnergy International, a Berlin-based consultancy specialised in decentralised energy services. The foundation was set through the Stanford Ignite Program in 2013, when I pitched the idea of swarm electrification for the first time, and the first business plan for SOLshare was developed. I was later joined by Hannes Kirchhoff, our CTO, and Daniel Ciganovich, our Director of Business Development, my two co-founders. In September 2015 SOLshare managed to install the world's first cyber-physical solar peer-to-peer sharing grid in a remote Bangladeshi village.

3.6.3 Scaling

After having set up our first 10 grids throughout the country, we recently received the US$1 million UN DESA Powering the Future We Want prize in collaboration with Grameen Shakti. With the help of this prize, we are determined to set the foundation to interconnect the maximum amount of Grameen Shakti's 1.7 million solar home systems in the upcoming years and also expand beyond the borders of Bangladesh, possibly even to Europe.

A major challenge here in Bangladesh is an aggressive approach for grid extension, indiscriminate of the associated costs. Electrification costs highly depend on the distance of how far the grid needs to expand to the next household and the associated load pattern. When we start interconnecting houses with cables, the closest house to the national grid could be the point of common coupling, that is, the one node that connects to the grid. In our R&D, we have built a Gridbox, which could potentially serve this purpose. The Gridbox is something like a Smart Metre or a bigger SOLbox, but with additional functionalities, such as converting the electricity from AC to DC and lowering the voltage. In addition, it could control the entire network connected to it. The government would then only have to connect to the point of common coupling. What is behind the cable infrastructure has already been developed by us. The only problem is that the tariffs of the SOLshare grid are significantly higher than the subsidised tariffs of the national grid. We have to do a comparative study on the costs of electrifying additional houses versus connecting just the point of common coupling and then paying the gap between the two tariffs to the connected households. That certainly depends on the distance, but it may be a commercially viable alternative. In Bangladesh, the range of DC appliances are being expanded every year. Light bulbs, TVs, fans, and now fridges are available in high-efficiency direct current. Water pumps, rice cutters, and corn shellers can all run on DC. What can the national grid offer these people that is better than if we managed to fully electrify them on DC and on solar home systems?

If we connected households with the point of common coupling, the national grid company would have to compensate us for building the local grid infrastructure. But this money does not have to come to us. It can go directly to the households and could even be a direct cash transfer.

We do not need the national grid. As our network can grow dynamically, we can incrementally add more capacity if we observe that there is more demand. We can then approach the households and suggest adding capacity, with which they can earn a lot of money. The entrepreneurial drive and financial incentive provided by our platform should be sufficiently strong to let the grid grow organically. From the feed-in perspective, the grid could be quite attractive, though. At times, it could be used as a buffer, and when it encounters problems, one could switch off all the connected households with just one node. It often occurs that when the grid is down and then starts running again, it immediately crashes because all households still have their switches on and the whole load affects the grid's stability. If the grid is down and then is switched on again, we can manage that gradually and thereby increase system stability.

The biggest challenge is getting aligned with the utilities and the government. The medium-term vision may depend on the aggressive rollout of the national grid. We plan our grid by giving people the financial incentive to make their local grid self-sufficient.

We can put our infrastructure on any system, including the low-cost modules of Chinese PV manufacturers. The appliances for the solar home systems are available everywhere in Bangladesh. Of course, if we build it on a poor-quality system, the question is: How long will it last? It is half the cost of a system provided by the IDCOL. It is available commercially and quick to install. One would not get any warranty. A recent World Resources Institute study showed that prices in Bangladesh are already very low compared to other countries, especially in East Africa, and the battery comes with a five-year warranty.

3.6.4 *International expansion*

Since winning a couple of awards with our concept in 2016 and 2017, we have been receiving numerous requests from all over the world. In principle, we are interested – but not right now. First, we have to carve out a solid business model here in Bangladesh: For the SOLcontrol it is done, but for the SOLbox it has not yet fully proven itself. Once we have that, we can start expanding to other countries. Of course, we have to take other factors such as population density into account. Would it then also be a B2B model, or rather B2C? Is there already a network of solar home systems that we can piggyback on, or do we need to start in a greenfield environment?

Cooperations and new technology developments

Our financial partner for the money wallet is bKash – the largest mobile money provider in the world, among others – which has recently surpassed M-Kopa in transaction volumes. Mobile money means that customers have their mobile money on their phones. bKash agents are present all across the country and have small stores where customers can charge their phones and make payments. We set up one wallet for these people, which we call the 'energy wallet'. Some people also call our system 'the energy bank'. One time a rural woman approached us and said that she really liked our system, but what she was missing was knowing how much money (!) was in her battery. That is a typical asset calculation approach. It is amazing to see how people in remote areas of Bangladesh have now started to realise that electricity now equals money for them. In urban Dhaka, we are far from that.

Recently, we have started to look into blockchain technology and the degree to which we could reap the benefits. We are collaborating with freeelio and the German Blockchain Association to develop a tokenisation model. We are setting up a programme label: Energy Efficiency in Germany for Energy Access in Bangladesh, where utilities or private consumers can decide to transfer their money from electricity savings directly onto the metre of the Smart Metres in the SOLgrids in Bangladesh. This will become the most transparent, efficient, and safe way to make a conditional cash transfer to a developing country.

3.6.5 *Market outlook and competitive environment*

In Bangladesh, the rollout of the national grid was very aggressive. We did a study of 350 households in rural areas: Half of them were connected to the national grid and the other half were close by, but without access to the grid. The result was that the solar home system provided a significantly better electricity service than the national grid, based on the multi-peer framework for measuring energy access, as developed by the World Bank. Quality of access here is measured against multiple attributes such as capacity, affordability, safety, reliability, and so forth. The national grid performed so poorly because of its reliability. The key indicator was the number of hours of available electricity between 6 pm and 10 pm. During these hours, the national grid often suffered from load-shedding in the rural areas investigated. What we observed on the ground is that if politicians promise households that they will be connected to the grid very soon, it is close to impossible to convince them to consider any alternative form of electricity supply. By contrast, with households that are already connected to the national grid and have had their experiences with it, we have a good chance because they know they need a backup – or even a complementary – system.

3.6.6 Interviewee biography

Dr Sebastian Groh is a 2013 Stanford Ignite Fellow from the Stanford Graduate School of Business (USA) and holds a PhD from Aalborg University (Denmark) and the Postgraduate School Microenergy Systems at the Technische Universität Berlin, where he wrote his thesis on the role of energy in development processes, energy poverty, and technical innovations. He has published a book and multiple journal articles on the topic of decentralised electrification in the Global South. Since 2014, Dr Groh has been working as the CEO and co-founder of SOLshare Ltd. and is an assistant professor at the Business School of North South University in Dhaka (Bangladesh).

On behalf of SOLshare, he received the Intersolar Award for Outstanding Solar Businesses, the UN Momentum for Change Award, both in 2016, as well as the 2017 Start-Up Energy Transition Award from the German Energy Agency (DENA) and the 2017 UN DESA Powering the Future We Want US$1 million Energy Grant, along with Grameen Shakti from Bangladesh. Dr Groh was further selected for the SE100 2017, a list of the top 100 Social Enablers around the world.

3.6.7 References

SOLshare. (2018), Create a network. Share electricity. Brighten the future. Retrieved February 10, 2018, from https://www.me-solshare.com/

3.7 Mobisol: developing a pioneering business model for off-grid energy in East Africa by starting with the users

Author: Klara Lindner, Mobisol GmbH

'A typical household in Bangladesh using kerosene lamps and rechargeable car batteries for lighting, TV, and other applications spends an astonishing €1.50 per kWh – compared to the subsidised price for grid electricity of only €0.03 per kWh in Bangladesh and the slightly higher price of more than €0.20 in Germany'.[54]

3.7.1 Context and origin of the idea

Although the electricity plug has long become indispensable in our daily lives, a quarter of the world still lives in areas without any form of modern energy provision – with most of those affected residing in sub-Saharan Africa (620 million). Countries such as Tanzania have electrification rates of 24 per cent. In rural areas, this number drops to less than 10 per cent. With the majority of the population living in village communities, this leaves 36 million people living off-grid in Tanzania.

In these unelectrified areas, only a few low-power-consuming activities are possible, and they are accessible only by people who own a diesel generator (and can handle its noise, fumes, and maintenance needs). In countries close to the equator, the sun sets at 6 pm, meaning it is completely dark at 6:05 pm. People who want to finish things thereafter, be it work, household chores, or homework, need to rely on candles, kerosene, or battery-run torches.

The majority of people living in these off-grid regions belong to the 'base of the economic pyramid' and have limited disposable income. In addition, the weak infrastructure in these areas – most of them are difficult to reach – makes this part of the population unattractive as potential customers for most companies around the world.

This is where Thomas Gottschalk comes into play. Driving around the globe with the 'Solar Taxi' in 2009, he noticed two things in the developing world. First, these people's realities are far from the stereotypes we are commonly confronted with in Europe: People rarely live from subsistence farming only – there are flourishing informal markets everywhere, especially in those areas that have some kind of grid connection, thus allowing for diversified

[54] Source: https://www.microfinancegateway.org/sites/default/files/mfg-en-paper-fact-sheet-the-potential-of-linking-microfinance-energy-supply-mar-2010.pdf.

economic activities. Second, the mobile networks in most rural areas were better than in the village of Gottschalk's grandmother back in Germany. And *every* person he met had a mobile phone, which they used not only for communication, but – in more and more countries – also for transferring money from person to person. Apart from that, as he was trained as an energy engineer, he knew of the price decreases in photovoltaic (PV) systems and how these, with their modular set-ups, are applicable in hard-to-reach areas. He sought out a handful of other like-minded people who had come to very similar conclusions. Together, we started to turn these three insights into a business: Mobisol.

3.7.2 Phase 1 (pilot phase): what do we provide then, really?

The basic concept was thought up quickly: Sell self-sufficient photovoltaic systems to individual households in off-grid areas, let customers pay with mobile money over time, and remotely turn the systems on or off through a modem with a SIM card that is placed in each system. But what we did not know was: What is the concrete value proposition, how do we reach our customers, and what does the detailed revenue model look like?

We found an angel investor who was fond enough of the idea to make a small grant for us to get started. Rather than wasting time on a 50 page investor deck, we turned the concept into a tangible prototype. At this point, it was basically a light bulb and a solar panel connected to a microchip with an embedded SIM card. Sized to fit in a suitcase, we could take it anywhere we wanted and easily demonstrate the concept to potential partners: Give them the phone number of the system, and by sending an SMS with the text '1', they would turn on the light and the penny dropped.

We first used this to get in touch with mobile network operators that were offering mobile money in off-grid areas and to create a service agreement with them. By that time, only four countries worldwide – Tanzania, Kenya, Ghana, and India – had implemented M-Pesa, the mobile money platform that radically eases the collection of small payments in rural areas. India was not included because we did not know anyone there, and the time zone difference would have made things even harder. So in May 2011, a small team of three set out on a field trip to Tanzania, Kenya, and Ghana with two aims: first, find a local, like-minded organisation to work with in rural areas; second, and most important, go out and talk to future customers, understand the way they live, the way they earn a living, and the role that energy plays in their lives.

We hired a car, found a translator, and started the engine. For the next four weeks, we would meet local renewable associations, drink tea in Tanzanian living rooms, attend weddings in Ghana, and hold discussions with the elders

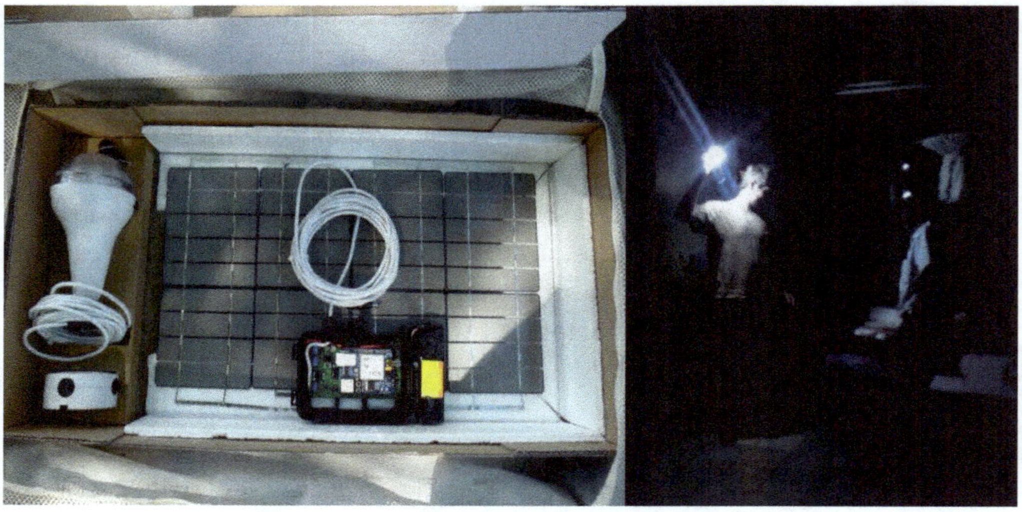

Figure 30: The suitcase-sized prototype and its demonstration to a potential partner.

Source: Mobisol/Lindner (2017).

of the Maasai tribes under the community tree. We quickly realised that the awareness about PV solar was high, but also that bad quality materials had ruined its reputation. People valued bright lights, but kerosene lamps were still 'good enough'. Getting access to 'real electricity' that powered stereos and TVs was what people really wanted. If there was something that they really wanted, they would find a way to pay for it. We realised that their ability to pay was much more volatile than we thought.

Based on this field research, we knew what the product of the pilot phase should look like: complete all-in-one kits big enough for real appliances (to fulfil willingness to pay), plus credit financed over three years (to ensure ability to pay). We wanted to include after-sales (to ensure long life), and favoured ownership rather than rental solutions (to make sure people take good care of the product).

So we designed a first offer that we could test with real customers: We went for three different PV all-in-one systems (60 Wp, 120 Wp, 200 Wp) and a 36 month credit and technical services agreement.

For this, a second hardware prototype was developed, now fully functional. We could remotely turn the PV system on and off based on incoming mobile money payments, and we could gather performance and usage data in real time to foresee maintenance activities through the modem inside the system. We used material from the Tanzanian DIY store for the casing, which was not only cheap, but ugly enough so that our pilot phase customers would understand that this really was only a first shot and that we would warmly appreciate their feedback.

Figure 31: The Mobisol Akademie in Arusha, Tanzania, and a trained installation technician in action.

Source: Mobisol/Lindner (2017).

With our local partner organisation, we created a concept for awareness-creation, marketing, and sales within a few months and started our pilot phase with 100 paying customers in October 2011 close to Arusha, Tanzania.

3.7.3 Phase 2 (becoming operational): how do we sell?

The product itself was clear now, but the remaining question was how to realise the business model in a viable way? What are the key activities and key resources on our side, and what do key partners do? The most challenging parts were: How to organise the high-quality and cost-efficient installation and maintenance of the PV systems, and how to distribute these efficiently in rural East Africa?

Meanwhile, we had also found a pilot-phase partner in Kenya and decided to run a second pilot with another 100 customers to test different service approaches with them.

Ensuring proper system installation and after-sales service

Since we had all been trained as engineers, we did the system installation ourselves in the beginning, but we quickly grasped that this would not be something feasible for a commercial roll-out. Our first move was to develop a plug and play kit that customers could install themselves. In co-creative sessions with real customers, we even managed to draft an accompanying manual. But we came to realise that even though our customers were now able to do the installation themselves, *they simply did not want to*. 'This is like having to put the engine into your new Mercedes-Benz!' is a quote from one of our pilot customers. People rather wanted someone knowledgeable to do the job.

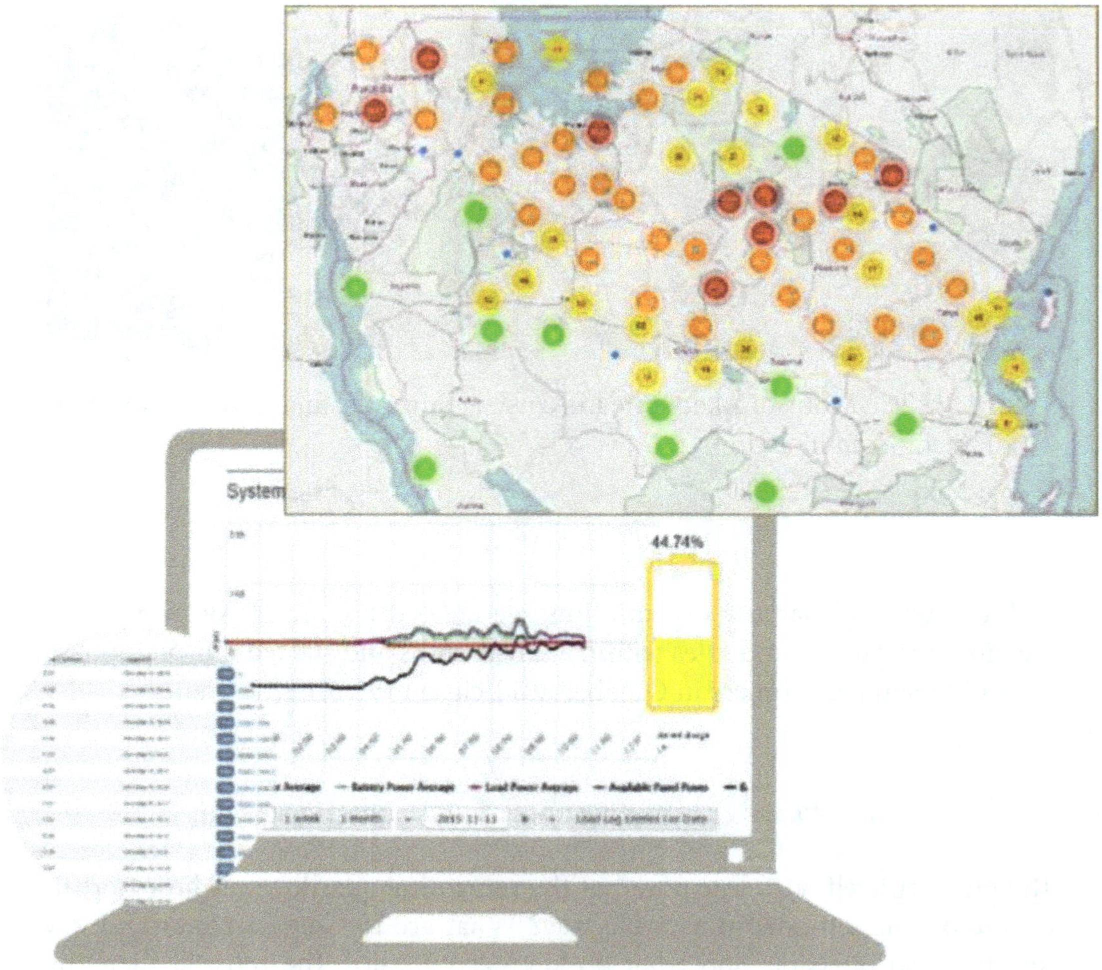

Figure 32: Mobisol's maintenance interface.
Source: Mobisol/Lindner (2017).

So we changed the approach and decided to bring local village technicians (who until then had been repairing houses, bikes, or phones) into the model. Through a two-week programme at the 'Mobisol Akademie', which we created for this purpose, they were trained and certified as Mobisol Installation Technicians.

After gaining some experience, the technician could later take another two-week training course and become a maintenance technician. The real-time performance and usage indicators from each system fuel an online database, which is our backbone for maintenance. A web-based interface makes it possible to coordinate technical activities in the village in an effective and viable way.

With that concept, we not only arrived at a feasible solution for us in the end, but we also made our customers even happier (they know and trust their local technicians more than foreign people) – and it created jobs in the village.

Figure 33: Mobisol's distribution centre and local promoters.
Source: Mobisol/Lindner (2017).

Making sure that our customers gain access, no matter how remote their homes are

To come up with a cost-efficient distribution strategy, we looked at informal markets for inspiration: When a Tanzanian family builds their new home, they go to the nearest market, buy the bricks, and find a means for transporting the materials home – sometimes a bus on the right route, sometimes a car, sometimes a boat. Every village then has a mason, who is hired to build the house.

We already had the local technician, so we started to build up a decentralised network of sales outlets. Today, we have our MobiShops at marketplaces that our customers regularly visit and make sure the packaging is optimised for easy transport. As soon as the papers are signed, our customers go out and cover the last mile themselves.

All sales activities are coordinated and monitored using a CRM system we developed in-house. Not only our customers use M-Pesa: We also deliver all our payments to staff digitally and thus have fewer transaction costs, less corruption, and a viable business.

3.7.4 Phase 3 (commercial roll-out): how do we finance our growth?

During our pilot phase, we had already dropped the idea that Mobisol could become a manufacturer or perhaps a wholesaler from which local distributors could buy in bulk and assume the point of contact for the end-user – these local distributors did not exist at the scale we needed. If we wanted to make this work, we thus had to extend our activities and build up our own structures. So the major remaining challenge was finding a way to pre-finance the hardware.

We wanted to work independent of subsidies or donor funds with a limited project life. We were aiming for a full commercial model.

Microfinance institutions seemed to be interesting key partners, but we quickly realised that, at least in sub-Saharan Africa, they only operated in cities and not in rural areas, where we envisioned our market. That meant we had to talk investors and banks into lending us a lot of money directly so that we could bridge those three years between the purchase and completion of payment.

We already had the switch-off mechanism that helped lenders to sleep better, but we still needed a way to carefully assess the creditworthiness of interested households. How do you do that without the equivalent of a *Schufa* credit rating and with a dynamic ability to pay? In several iterations, we came up with a credit survey that carefully assesses all potential income sources on the one hand, and all expenses on the other. Surprisingly, applicants have two to three income sources on average, and one of the most important questions to understand the expenses of a household was 'How many wives do you have?' Over time, we developed a double-check algorithm from the data we had gathered that allowed us to see who was trying to tweak the responses.

As the business evolved, we 'walked up the finance ladder': from angel investors and foundations in the beginning, to donor funds and impact investors, to finally becoming bankable and getting 'ordinary bank loans' from KfW Group, the German government-owned development bank, and the like.

3.7.5 Phase 4: how to become a market leader?

The year 2012 was all about piloting, prototyping, and iterating the business model. Then, in 2013, we sold 2500 systems in our Tanzanian market. In 2014 we opened Rwanda and sold 10,000 in total. In 2015 we started to scale and reached 25,000 households. In June 2016, we reached the symbolic moment of empowering household number 50,000 and by then had become one of the three major players in off-grid electrification in sub-Saharan Africa (the other two were M-Kopa and Offgrid Electric).

Currently, we are facing two new challenges: We are aiming to cross the 'chasm' of innovation between early adopters and mainstream customers, and we want to continue our regional expansion without having to simultaneously grow the company as well.

In contrast to markets of the Global North, so-called below-the-line marketing plays a much more important role than do above-the-line activities in our context (direct communication vs. TV and radio ads). Through our 'Project Saturation' we are currently developing marketing strategies tailored to different customer types and building long-term relationships with trust-building entities and influencers, such as schools, health care facilities, and religious communities.

In parallel to this vertical growth, our regional expansion team is aiming at finding the right partners to start joint ventures in new regions. Our main lesson learnt is that having experience with renewable energy technologies is not an important criterion. What is much more crucial is for the partner to share a similar mindset of engaging in a long-term – and at the same time commercial – relationship with customers. For example, a motorcycle vendor

Figure 34: A Mobisol marketing officer meeting potential customers at a Maasai market in Tanzania.

Source: Mobisol/Lindner (2017).

that is active nation-wide could be our next partner. By developing a 'Mobisol Blueprint' or operational manual, we aim to bring that partner up to a working level quickly.

The outcome so far is that we are now selling about 4000 systems per month. After Rwanda, we are now starting operations in Kenya in a joint venture with a national car (and solar) battery distributor and have pilot systems set up in two more countries.

3.7.6 Summary – finding new ways to serve the underserved

For us, the so-called developing world is no barren land but a green field for innovation. Mobile phones, mobile banking, and renewable energy technologies are only the start, and we see a potential for leapfrogging in many more areas.

A few things we are working on at the moment

Seeing the radical uptake of smartphones, we have also created a customer app that increasingly serves as a lean interface between us and the customer. As we have gained a strong reputation for good quality and generally 'being there' in the village, we are starting to create a position as a 'gateway' between our customers and other product and service providers and are currently testing that with electrical appliances and health insurance. Our most radical R&D project is targeted at overcoming the bad infrastructure in rural Africa. We

Figure 35: Mobisol drone landing on a customer house.
Source: Mobisol/Lindner (2017).

are currently testing a drone delivery network for spare parts, piggybacking on the ever-growing network of customer homes, which could serve as battery-recharging stations.

3.7.7 Author biography

Klara Lindner strives to connect human-centred design with sustainable energy provision. She joined the solar company Mobisol in its infancy, led the pilot phase in East Africa, and co-developed its pioneering business model. Alongside improving Mobisol's customer experience, Klara became part of the research program Microenergy Systems in 2013, investigating service design in the bottom-of the-pyramid/energy context. As a certified Design Thinking Coach, Klara has been using various workshop settings to teach creative thinking applicable to processes of innovation and change.

3.8 Solarkiosk: social enterprise and decentralised energy

Interview with Lars Krückeberg, Founding Partner of Solarkiosk, on January 18, 2016

Solarkiosk enables and empowers the sustainable economic development of base-of-the-pyramid (BoP) communities world-wide through the provision of clean energy services, quality products, and sustainable solutions. Intertwining an award-winning technology solution with an inclusive business model, Solarkiosk fosters local entrepreneurship at the BoP. The first Solarkiosk project was successfully implemented in 2012. By the beginning of 2018, Solarkiosk had established six country subsidiaries and is involved with projects on three continents (Solarkiosk 2018).

3.8.1 Technology and business model

Technology

In our profession as architects, my partners and I are always keen on getting to know new, holistic approaches that step beyond the day-to-day business of an architecture workshop. The idea of providing decentralised energy via a kiosk resulted from the observation that informal markets exist all over sub-Saharan Africa, including areas beyond the grid, and kiosks are a familiar feature even in the most remote regions.

Typically, development projects in these regions include the construction of mini-grids, providing power to schools, selling solar lamps, etc. The organisations provide the hardware and then leave. As architects, we deal with energy from a perspective of sustainability – energy needed to construct a building, to operate and maintain it, to produce the materials needed for its construction, etc.

We looked at the idea of a kiosk from a purely technical point of view: What type of – not too heavy – building could be manufactured on an industrial scale, such that it could be transported anywhere, erected very easily, and, once erected, could produce power in a safe and sustainable manner while requiring low levels of maintenance? You have to imagine it as a 'power room' in which other commercial operations are also possible. We talk about last-mile distribution, both in retail and electricity production. This last mile is tricky, and that is why so few ventures have actually succeeded.

The idea was to develop a product that can adapt to any climate, can deliver clean energy in a sustainable way, is easy to transport, and is modular. That means it should be able to connect to other types of energy supply, such as wind or biomass, but it should also be extendable in a spatial dimension.

Figure 36: On-site construction of a Solarkiosk.
Source: Solarkiosk (2017).

We strive for an integrated solution in which architecture and solar power are so closely tied together that human error can be minimised.

Business model

We always had the vision that we wanted to have an impact, but a lasting impact can only be achieved if it is linked to a successful business model. We call it social business.

When we started, we perceived the world as being divided into two factions that have only since, say, 2014 converged: the business sector and the social sector. For the social sector, the premise existed that a project was not supposed to make a profit because it required ripping off the poor (although NGOs are also surviving on that). By contrast, the business sector does not care about social impact because it simply costs too much. The financial flows were separated accordingly.

It was very difficult to make people understand that, in this area, the combination of the two worlds is the only implementable option that allows for scalability. Many NGOs initiate great projects, but they are often hard to scale, and their financial sustainability is rarely achieved.

With a business approach, the social impact may not be the only focus. However, the projects reach acceptance levels that have a positive feedback effect on performance and, hence, the continuation and expansion of the project.

Meanwhile, we started developing the business model for the kiosk – also based on the principle of modularity – and attracted the attention of one major investor who is active in solar energy. This investor allowed us to do a proof of concept.

Anyone who is familiar with the context of sub-Saharan Africa immediately realises that there is an immense business opportunity. These markets are very dynamic. Companies such as Mobisol and M-Kopa sell energy solutions that range from small lamps to proper solar home systems. The problem is that rural residents typically do not have any money and have to pay in instalments. The key to this business model lies in the financing system. M-Kopa is comparable to mobile banking service M-Pesa. There are currently deals being undertaken in the range of hundreds of millions of US dollars.

Similar to the kiosk, our business model is equally flexible. One has to gradually explore all the possibilities linked to it.

3.8.2 History and organisation

In Berlin, we have around 30 employees. Together with our six African affiliates, we now have a workforce of more than 100 people. Our subsidiaries are all locally registered companies.

It all began with a meeting between my partner, Wolfram Putz, and Andreas Spieß, who developed the idea of providing decentralised energy via a kiosk after making the observation that informal markets exist all over sub-Saharan Africa. Together with Wolfram Putz and Thomas Willemeit, we are the founders of the architectural practice GRAFT, with subsidiaries on several continents. We are currently building a children's hospital in Ethiopia and have been frequently visiting Ethiopia, which is our first project in sub-Saharan Africa. There we met Andreas Spieß, a lawyer from Berlin, who had founded a solar company called Solar23 in Addis Ababa in 2008, which is now one of the biggest solar system integrators in Ethiopia. Initially, Solar23 was a spinoff of Siemens: When Siemens closed down its operations in solar generation, some of its employees became entrepreneurs.

We started with Ethiopia, which is a fascinating but difficult and very bureaucratic market environment with high barriers to entry, especially for founding new companies. The population is very poor with limited buying power, even if they desperately need the electricity. Our assumption was that the poor would redirect the money that they would typically spend for 'dirty' energy, such as kerosene, diesel, or paraffin, to our cleaner and cheaper energy. No one in the world spends a higher proportion of their income for energy than those deprived people because these forms of energy – in particular diesel – become very expensive when they reach these remote areas. A cleaner source of energy has immediate influences on their lives because it makes them healthier and saves them money. The idea is that these people will slowly climb up the 'energy ladder'. In the beginning, a household may only have the financial resources to buy a pocket lamp, then comes a small home system, then a larger system, and so forth – up to the replacement of the diesel generator. The same is true not only for residential customers, but also for small businesses.

In Ethiopia and in Kenya, we delivered the proof that the concept actually works. We placed our kiosks literally at the end of the world to see whether people in these areas would accept them because there was no available data on such an endeavour. These are informal, but also untapped, markets. However, if you place half a dozen kiosks at the end of the world, the business is highly unprofitable because the supply chain is too expensive.

We tested in Ethiopia what would work and what would not work. We found the sites and operators and talked to the communities to get to know what locals really needed in order to create awareness, which was the task of our subsidiaries. We support them in marketing and give hints about best practices in other countries, but we learn from our operators and the agents in the field on a daily basis. Our local teams visit the kiosks up to one time per week to learn. However, scaling our business model, adding corporate functions and compliance mechanisms, optimising the logistics, etc., is all undertaken from our headquarters.

We started with the assumption that we have to convince people to come to the kiosk, which means we need traffic. So we have to offer something that people need. Once they have arrived at the kiosk, they realise there is light at night and see that solar power actually works. Meanwhile, we can offer cold drinks, play music, and provide a social space. We opted for fast-moving consumer goods – one pillar of our sales strategy. In addition, we offer special products related to hygiene. The second range of products relates to solar energy. That is where the impact actually starts! From a pocket lamp to a full-fledged home lighting system, we provide energy solutions for the residents.

The major problem is awareness: People will only buy what they are familiar with. Unfortunately, a first wave of cheap, low-quality products from China had destroyed confidence in solar products because they failed very rapidly.

The third business line for our kiosks is energy services. When the kiosk generates electricity to operate its point of sale – including having the lights, a small computer, and a solar fridge on – the kiosk can sell its excess production to other businesses connected to the kiosk. Alternatively, we can use the excess electricity for entertainment, for example for TV. The TV can then be transformed into a small cinema, and the operator of the kiosk can charge an entry fee to show football. We provide mobile phone banking, phone charging, internet services, and much more.

We had to take into account the need to achieve community acceptance: Who are the stakeholders in the community? What do the people in a particular community actually need? What are they interested in? And how to

Figure 37: A Solarkiosk in Botswana.
Source: Solarkiosk (2017).

create awareness? We had to start becoming retailers as well as experts in marketing and things we never dealt with before. We realised that we could not start at the end of the world, but rather that we had to expand outwards in rings and clusters.

We then founded a joint venture with a family in Botswana that operates one of the largest franchise chains in sub-Saharan Africa – real retail professionals. They showed us how retail in Africa is functioning. The first thing that they decided was that all goods and products that are delivered to – and sold at – a kiosk pass via the point of sale; everything that comes in and goes out is controlled. In fact, a kiosk cannot be successful without this control, otherwise there may be theft. We sharpened our understanding of how to maintain close ties with our operators. We do not employ them – we would rather create a partnership with them and formulate precise contracts. We would not sell our kiosks because they would be too expensive. However, we would enable local entrepreneurs to start a business with it and create an environment where other businesses would also flourish in order to create an impact on the entire community.

We discovered that the platform, that is, the kiosk, has a certain value to others – and they are willing to buy it. This became our second pillar for revenues – a cross-financing tool, if you will – and enabled us to finance our expansion, which is quite capital-intensive, as well as improvements to the system.

Wherever a kiosk is established, other businesses emerge, too. We started to introduce fast-food services. Our clean-cooking stoves have a fundamentally positive health and environmental impact, in particular with regards to deforestation. The operators will pay us rent for the equipment; they are like a franchise.

Now we are learning how a kiosk is the nucleus and trigger for other ventures, such as rental space, a cinema, and connectivity. We just signed a partnership with SES, the largest satellite provider in the world, headquartered in Luxemburg. With their support, we can bring the internet to the most remote places in the world. Of course, that is more expensive than a landline, but one of the criteria for the locations of our kiosks is the very absence of functioning transport, energy, and telecommunications infrastructure. In those areas, quantum leaps are possible, as telecommunications has demonstrated. It is not astonishing that Google, Amazon, Richard Branson, and Elon Musk have spent billions of dollars to reach these remote markets. It is a positive development to provide these people access to goods and information. But it is also a gigantic business opportunity. For that type of infrastructure, a person in charge has to be on site to market and sell this offer. People then also need devices to access the internet, and these devices have to be charged. When spending all these billions in space technology, investors expect these services to emerge automatically, but that is not the case. Rather, they have to be initiated by organisations such as Solarkiosk. We are an analogue road to market, but we pave the digital road to market.

A kiosk typically creates about four new jobs. Our operators start earning money from day one of the operation of the kiosk. But we want to break even jointly for all the kiosks in a country, and also for our headquarters in Berlin. That is only possible via scale, whereby we can increase our buying power, accommodate the high capital expenditure, and seize other opportunities because of our position as a monopoly provider. Our business plan is to break even within the next five years, but we have to build more than a thousand

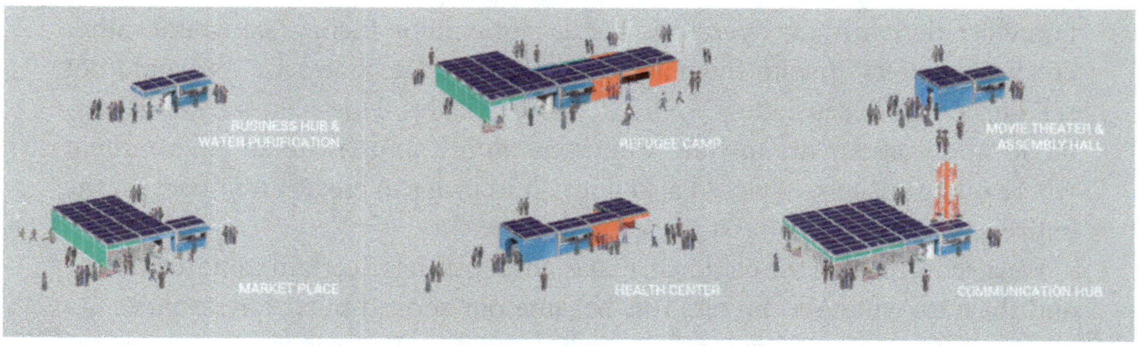

Figure 38: Potential use cases of a Solarkiosk.

Source: Solarkiosk (2017).

kiosks to reach that target. To do so, you have to acquire the finance and find the people who believe in our venture.

3.8.3 Scaling and cooperations

We first worked together with a US software company, but we did not like that the data was being stored in the United States, and it was not available in all countries where we wanted to operate. We then decided to purchase software developed in the respective countries, but the software was mostly targeted towards supermarkets as customers. For the informal market, solutions did not exist. People just counted the money themselves. So we came to the conclusion that we had to develop our own software and purchase the corresponding hardware.

We went to international conferences and presented our model. At that time, we had around 15 kiosks in three countries in operation. We soon realised that interested people were approaching us because we had a unique selling proposition.

The European Union, USAID, and the World Bank started to allow for private-sector involvement in their projects or to explicitly integrate it. For example, the EU's Electrify programme started to steer major financial resources towards the private sector – money that, in previous times, would have been exclusively reserved for NGOs. The decision makers redefined their ideas about social enterprise. Large investment funds were approaching us, too. At the same time, representatives of multinational corporations were approaching us. They were either from the fast-moving consumer goods sector, or they were companies from the energy sector with a profound knowledge of the African business context, namely Coca-Cola and Total.

In order to secure their future markets, Coca-Cola tried to develop something similar to our idea. They called it EKOCENTER, but it did not work. It also evolved from the idea of a social enterprise – a triangular relationship between government, business, and NGOs. This idea is not in the context of corporate social responsibility, but rather a profitable business that enables and empowers people. They painted a large container in red and added a massive number of gadgets and functions. In August 2013, their CEO and Chairman, Muhtar Kent, announced the launch of the first EKOCENTER in South Africa. Soon afterwards, they realised that it was just way too expensive. They actively searched for alternative solutions and found us: 'If you can't beat them, join them!' So we did a pilot with 25 kiosks with them in countries where we were already operating, as well as three new countries. The pilot was a success. Then we started phase 2 with 150 kiosks, which is ongoing. In the meantime, they have seriously started investing in our organisation and venture, and they believe in it.

The cooperation was based on the following agreement: We own the assets, we put them on site, we operate them and do the business. But EKOCENTER is present for all decisions, we decide together on how to improve the operations, and the kiosks are branded 'EKOCENTER powered by Solarkiosk'. For them, it is important to establish their EKOCENTER brand. Of course, we sell Coke in these kiosks. They are interested in developing village concepts with a focus on the social dimension by electrifying schools and medical dispensaries as well as a particular focus on water purification – a powerful lever for the physical and economic health of a continent. We could do that on our own – we even founded a charity together with the Siemens Foundation that focusses on water purification in Africa: the Solar Fountain GmbH. However, I do not personally believe that it is a viable business model. I am rather convinced that water supply should not be a commercial service to make a profit. Water is essential for survival, but electricity is not.

The other business partner that approached us early on was Total. We launched two pilot kiosks with them in Kenya. In the meantime, we have sold more kiosks to Total. We also sold 18 kiosks to Coca-Cola. There are some regions where we see no business potential but our partners do. There is exclusivity – maybe one day we will decide to tackle these regions, too. But for the moment, we just install the kiosks there and then leave the operations to our partners.

At the end of 2016, we had around 200 kiosks that are operated by us in six different countries. From the other kiosks, we can still learn. We are still in the learning process about cost optimisation and the lean management of retail space. The supply chain logistics are still a major bottleneck. The other one is

Figure 39: Coca-Cola's EKOCENTER.

Source: The Coca-Cola Company (2017).

human capacity: Where do I find the right operators in Africa who can deal with the retail challenges, the marketing and technological challenges, as well as the real-estate dimension at the same time?

More recently, we started a pilot project with Coca-Cola and Ericsson in Rwanda. We provide energy and rent out the area for a telecommunications tower. In return, we pay for their internet service provider and the data used. These are complementary services. Customers who come to eat and drink at the kiosk may also want to use the internet. These types of experimental settings could be called 'de-risking'. The bottleneck is always the human factor: Which operators are capable of handling such a complex service? How to find entrepreneurs who are both loyal and realise their own ideas on top of our business, ideally women? They have to be sales agents, and sometimes they have to employ additional staff. Finding the right site for the location of the kiosk is equally challenging: even the best entrepreneur fails if the location is not carefully chosen because no traffic emerges. It is also difficult to transport the kiosk into the countries, so we have set up local manufacturing units. Often the prices are lower when components are imported than if they are produced locally. But it is essential to have the option of producing components locally.

We started with Ethiopia, Kenya, and Botswana. Later we added Rwanda, Tanzania, and Ghana. There are also kiosks now in Vietnam. We won an EU tender and are now building 40 kiosks in Ethiopia and 40 in Kenya with EU funding and using a local manufacturing workforce.

We started developing products for Connected Solar Clinics and Connected Solar Schools, in particular for governments and the United Nations. In early 2016, we launched a Connected Solar School in the Jordanian refugee camp Zaatari, together with SES. We donated the Solar School, but we are convinced that there will be business opportunities in the future. Energy access for refugees is essential, but connectivity is equally important, especially for education. The Connected Solar Clinic was built in the Jordanian region of Al-Mafraq, which hosts many Syrian refugees. Up to 30 per cent of the population there consists of refugees. We want to show that energy and connectivity are a solution for – let's face it – the new cities of the world. That type of infrastructure could potentially be built in any slum or informal settlement around the world. It is particularly useful if water purification is added.

We offer a piece of infrastructure for the cities of tomorrow because it can become the nucleus of urban development due to its modular design. It can become a mini-shopping mall, a place for assemblies, even a security post because of the connectivity. There are an infinite amount of possibilities and synergies. We can imagine building a complete neighbourhood around a kiosk. Our advantage is that we provide the nucleus for clusters of businesses and ideas. We provide the platform to create new jobs and promote the local economy.

Of course, we do not want to stay in Africa. We plan to expand to Latin America and Asia, but we have to proceed step by step. We would not survive by becoming too big too fast. We first have to develop deeper roots in the territories where we are operating.

As soon as we are able to enter the wholesale business, everything will change.

3.8.4 Market outlook and competitive environment

One of our main competitive advantages is that we have successfully established the infrastructure and matrix organisation within the different countries. No competitor is able to replicate these structures so easily. Especially China has a strong strategic interest in Africa, but it tends to lack the soft skills. The confidence of the communities is the key to success. That is why the multinationals turn to us, and without them it would not work. We would not prevent anyone from imitating us, but the model is very tricky in its implementation.

3.8.5 Interviewee biography

Lars Krückeberg, M.Arch, Dipl.-Ing. Arch, Architekt BDA, Founding Partner of GRAFT

Lars Krückeberg studied architecture at the Technical University Braunschweig, Germany, the Universitá degli Studi di Firenze, Italy, and the German Institute for History of Art, Firenze, Italy. He graduated as Dipl.-Ing. Arch in Braunschweig and received his Master of Architecture at the Southern Californian Institute of Architecture SCI Arc, Los Angeles, USA.

In 1998 Krückeberg established GRAFT in Los Angeles together with Wolfram Putz and Thomas Willemeit. With additional offices in Berlin and Beijing, GRAFT has been commissioned to design and manage a wide range of projects in multiple disciplines and locations. GRAFT has won numerous national and international awards and earned an international reputation throughout its 15 year existence.

In 2009 he co-founded Solarkiosk GmbH together with Putz, Willemeit, and Andreas Spiess in Berlin and manages the company as acting CTO. Since 2012, affiliate companies have been incorporated in Ethiopia, Kenya, Botswana, Tanzania, Rwanda, and Ghana.

3.8.6 References

Solarkiosk. (2018), *SOLARKIOSK – enable. Empower.* Retrieved February 10, 2018, from http://solarkiosk.eu/company/

3.9 Power Ledger: peer-to-peer trading with Blockchain as decentralised transaction technology

Interview with Jemma Green, co-founder and chair of Power Ledger, on August 25, 2017

Power Ledger uses blockchain technology to enable households and buildings to trade excess renewable energy peer to peer to make power more distributable and sustainable for consumers. The Power Ledger system tracks the generation and consumption of all trading participants and settles energy trades on pre-determined terms and conditions in near real time.

In October 2017, Power Ledger raised A$34 million through its POWR Token Generation Event. More than 15,000 buyers took part in Australia's first Initial Coin Offering, with the Main Sale following a successful Pre-Sale at the start of September that saw the company raise A$17 million in just 72 hours (Power Ledger 2017).

3.9.1 Technology and business model

Technology

Of all blockchain start-ups in the energy sector, Power Ledger has advanced the peer-to-peer trading environment the most. Power Ledger facilitates peer-to-peer energy trading – a concept by which renewable energy can be sold between buyers and sellers. This is an innovative business model capable of disrupting incumbent utility companies. Power Ledger has a unique service offering, as we partner with utilities, allowing them to on-board their customers to the platform. We call it 'phased disruption', and the response from the market has been very positive.

As for the technology, Power Ledger utilises blockchain technology, which is a secure, immutable, and transparent database. Because the platform manages financial transactions, the security of the database is crucial. The ledger is distributed across many nodes (versions of the database that are continually shared and reconciled to reflect the same ledger). Blockchain technology is also cheaper and faster than traditional databases, which is necessary, as energy trading requires millions of transactions.

Our software, which is already live, connects to Smart Metres and brings the data into EcoChain, our proof-of-stake private blockchain.

For peer-to-peer trading, the platform is able to register and settle transactions. Within a block of flats, for example, the platform allocates electricity to each flat and enables residents to sell their excess electricity to their neighbours. Our two-token structure enables us to process payments in the resident's local currency rather than through cryptocurrencies. In this instance, people can load funds onto the platform, which is linked to a credit system called Sparkz. This enables consumers to use Sparkz to purchase electricity or for a producer to sell it. Then they are able to convert the Sparkz back to dollars or any other fiat currency[55] – customers do not need cryptocurrencies or Bitcoin at all. The only parties that need to access our POWR tokens directly are our application hosts.

The validation is performed via a proof-of-stake blockchain – it has a quick block time, can be used in any energy market, and was purpose-built for high-volume transactions.

We also have a product called Asset Germination Events, whereby consumers buy, hold, and trade fractional ownership of large renewable energy assets, such as batteries or solar farms, using our platform. For example, a solar farm could use the blockchain to issue ownership. If someone owns 3.7 per cent of that asset, then that person would receive 3.7 per cent of the revenue.

For the solar farm to connect to the grid, the operator would need a power purchase agreement, but by using the blockchain, they can sell to many more consumers. At the same time, a building might purchase electricity from their own solar farms down the road to manage that energy transaction.

Blockchain gives us the ability to transition to a low-cost, low-carbon, and resilient energy system. This makes the grid more resilient, because by using blockchain, we can create localised energy systems and enable consumers to access electricity from a diverse number of sources. Today, households with solar panels can sell their excess energy to a utility company, which then sells it to others. The energy goes back to the grid and may have to travel far before reaching its destination. Our platform allows solar panel owners to sell their energy to those closest to them, which means the energy does not have to travel as far.

With respect to data privacy, residents are not forced to participate in our scheme, but they could if they wanted to – and they would be remunerated for it. We had cases where consumers were offered A$25 to turn off their devices during peak demand. Of course, there are security concerns, but using blockchain technology mitigates this risk, as everything can be encrypted and anonymised. Blockchain is actually the solution to these concerns.

Business model

Power Ledger earns money via a daily fixed supply charge. We collect a small amount of revenue from each kilowatt hour sold. With the peer-to-peer trading,

---

[55] According to Investopedia, fiat money is currency that a government has declared to be legal tender, but it is not backed by a physical commodity.

there is also a premium on each kilowatt hour sold, like a transaction fee. The supply charge varies and depends on the daily fixed supply charge set by the retailers in the marketplace, so it is benchmarked against that. By contrast, the peer-to-peer trading charge depends on the volume of the market. For example, we charge lower fees for higher voltage.

The tenant in a flat pays the electricity bill to the landlord. If the owner has rooftop solar, it provides them with an income stream and a return on investment for the solar panels and the battery. A solar system offers a return on investment of around 25 per cent in Australia. But homeowners can also purchase differential power if they do not have sufficient solar power in their systems. Owners can also purchase electricity on the wholesale market and then sell it at retail cost to the tenants.

Australian legislation allows us to operate embedded networks within buildings and enables us to do trials for 1.5 per cent of the turnover of a company without an exemption. Thereafter, we can apply for exemptions from the regulator to conduct larger trials.

At the moment, we offer a flat rate for electricity. If the right pricing is in place – taking into account the time of day – a rooftop producer may be remunerated for it. Even when you are uploading solar power during the peak period, you would not be adequately compensated.

3.9.2 History and organisation

In May 2017, we had nine staff members in our company, including blockchain developers and experts. Since that time, we have grown to about 20 staff members, which includes energy experts, energy economists, and regulatory specialists. In addition, we have hired a business development team, a legal advisor specialising in blockchain issues, a chief operating officer, as well as added to our user interface and user experience (UX/UI) capabilities.

3.9.3 Scaling and cooperations

The first peer-to-peer trading project we did was in Busselton, south of Perth. That project ran from September to December 2016. The second project with peer-to-peer technology – a trial with 500 sites – was in Auckland with Vector Limited and started in December 2016. In March 2017, we started the Gen Y Housing Project, which is a block of flats in Perth. In addition, we have signed deals with BCPG, one of the largest solar Independent Power Producers in Thailand, Tech Mahindra in India. In Europe, the Liechtenstein Institute for Strategic Development (LISD) will become the first Application Host to offer Power Ledger's blockchain-based peer-to-peer energy trading platform in Europe, and we are part of a Smart Cities project in Fremantle, which is partially funded by the Australian government.

There are lots of opportunities in Asia and Africa. In developing countries, the Power Ledger platform can provide a modern, low-cost, low-carbon alternative to the traditional energy-supply model. Even better, because of the ability to fractionalise an asset, the platform can give communities and individuals a chance to own a share in their local power-generating assets. Regulated markets such as those in California, New York, continental Europe, the United Kingdom, Germany, Italy, and Austria would also be of interest to us.

For the most part, utilities are aware that something needs to change. We want the Power Ledger platform to disrupt the energy industry, not to destroy its value, so we are working closely with traditional energy providers to demonstrate new ways for them to stay relevant. Network operators need to find a way to keep people from defecting from the grid, and retailers need to find ways of connecting with their customers – so we are helping them achieve that.

The retailers we speak with realise that whether Power Ledger had come along or not, change would have happened – more renewables and grid defection, for example. There is also a lot of customer churn. One way for retailers to have a longer-lasting relationship with their customers is to offer peer-to-peer trading. It is effectively cannibalising their incumbent market, but if they do not do it, someone else will. I think they realise disruption is inevitable for them, but the suffering is optional.

Our relationship with these utilities is not tense, but sometimes apprehensive. You need to demonstrate and persuade 20 people, sometimes more, before you can move forward with a deal. We are creating a new market, which means we need to have a number of conversations, explain what the blockchain is, and get people comfortable with the company before moving forward. It is a long engagement process.

The retailer landscape is very diverse. We have not been met with any hostile responses, but it does seem that the smaller retailers have generally been more responsive. In addition, there are some IT and electric vehicle companies involved as well as universities.

3.9.4 Market outlook and competitive environment

Our regulator, AEMO, hired Audrey Zibelman, who was previously the chair of the New York Public Service Commission and who has done a lot of work to manage that transition from a centralised to a hybridised system. That kind of market reform needs to happen in order to enable the technology to be deployed, not just by disrupting the technology but by managing the system resiliently during the transition. There is quite a lot of market reform that needs to happen, but there seem to be positive sentiments about renewable energies, with a couple of exceptions. For example, South Australia, where they have a high penetration of rooftop solar systems, endured blackouts last year. It was a big political issue. The government said that renewables were to blame, but in my opinion, it was actually a problem with the software and interconnectors. The argument that politicians put forward was that we should have fewer renewables.

This resulted in Elon Musk offering to put a big battery into the South Australian grid to show how renewables could solve the issue. Musk spoke with our prime minister and the South Australian government. Now the government has announced that batteries will be part of the solution, and that it is not just turning on more gas-fired power stations. The other part of the discussion concerns carbon emissions. Companies can just go on polluting, which means it is not a level playing field. We really need an instrument such as a baseline credit scheme or a carbon tax to address this issue. Politically, it has become a hot potato, and no one wants to touch it, but it needs to be addressed, along with a reform of the electricity market.

3.9.5 Interviewee biography

Jemma Green, Chair of Power Ledger Pty Ltd, is a member of the board of directors

Dr Jemma Green has more than 15 years of experience in finance and risk advisory, having worked for 11 years in investment banking in London. Whilst there, she completed a master's degree and two postgraduate diplomas from Cambridge University.

Jemma is a research fellow at the Curtin University Sustainability Policy (CUSP) Institute. Her doctoral research into 'Citizen Utilities' has produced unique insights into the challenges and opportunities for the deployment of rooftop solar PV and battery storage within multi-unit developments and the application of the blockchain.

Jemma is also experienced in the challenges of sustainable cities through her role as an independent Councillor of the City of Perth.

3.9.6 References

Power Ledger. (2017), Power Ledger token generation event. Retrieved November 27, 2017, from https://powerledger.io/

3.10 Core competencies in the energy transition – insights for corporate and political decision makers

The conclusions of this chapter follow a dual structure. In the first part, the authors establish a categorisation of business models based on the conducted interviews and the contribution by Klara Lindner on Mobisol. In the second part, six core competencies for decision makers are derived from the interviews.

3.10.1 Three new business models for dealing with the energy transformation

Taxonomies of business models have seen widespread use to classify ways that companies generate money. One of the most holistic attempts is suggested by Gassmann et al. (2014). The authors claim that 55 business models are responsible for 90 per cent of the world's most successful businesses (ibid.), ranging from classical models such as franchising and multi-sided platforms to digital models such as freemium and crowdsourcing. By contrast, Hamwi and Lizarralde (2017) outline three major business models they observe in the energy transition: 'Customer-owned product-centered business models, where customers own the product related to the electricity generation or management; third-party service-centered business models, where a third party offers energy services to the customer; and energy community business models, where resources are pooled and shared between community members' (ibid.).

The seven interviews and contributions of the founders and executives of start-ups serve as an empirical basis to develop a taxonomy that bears similarities to what is described by Hamwi and Lizarralde (2017), albeit we chose a different terminology that takes into account the blend of regulated and non-regulated elements of the value chain, and its role as part of so-called 'critical infrastructure'.

The following three business models can be derived from the analysis:

- **New asset-ownership models:** the start-ups SOLshare and Mobisol enable private energy consumers to become self-producers. Via different financing schemes, their customers will eventually own the devices they are using.
- **New service and operating models:** the start-ups Envio Systems and Solarkiosk as well as Timo Leukefeld's energy-efficient buildings belong to this category. In all three cases, services to establish or improve energy use are offered to final customers or intermediaries. Envio Systems builds on the existing energy supply infrastructure and specificities of each commercial dwelling to increase its energy efficiency. Solarkiosk rents its kiosks to

local operators and lets them decide which services they want to offer to their customers. The tenants who occupy Leukefeld's single or multi-family houses pay rent to the building agencies, banks, or utilities that are financing and constructing the buildings.

- **New platform models:** for the first two business models, continuity can be observed, stretching from the first attempts of liberalisation to full-fledged decentralisation. The platform models, however, are an offspring of digitalisation, which allows for new markets of buyers and sellers of certain products and services. Entelios (in the area of Demand Response) and Power Ledger (in peer-to-peer trading) fulfil that role. They do not enable consumers to acquire assets, nor do they rent any devices or own any assets themselves. They only provide the intelligence to coordinate the assets that belong to their customers.

Combining the three phases of the energy transformation with the taxonomy of business models, Figure 40 shows the categorisation along the two dimensions.

In the following sections, the three types of business models are presented in greater detail and complemented by additional examples.

New asset-ownership models

Together with the water supply, transport, telecommunications, and waste management sectors, the energy sector – in particular, the grid-based infrastructure

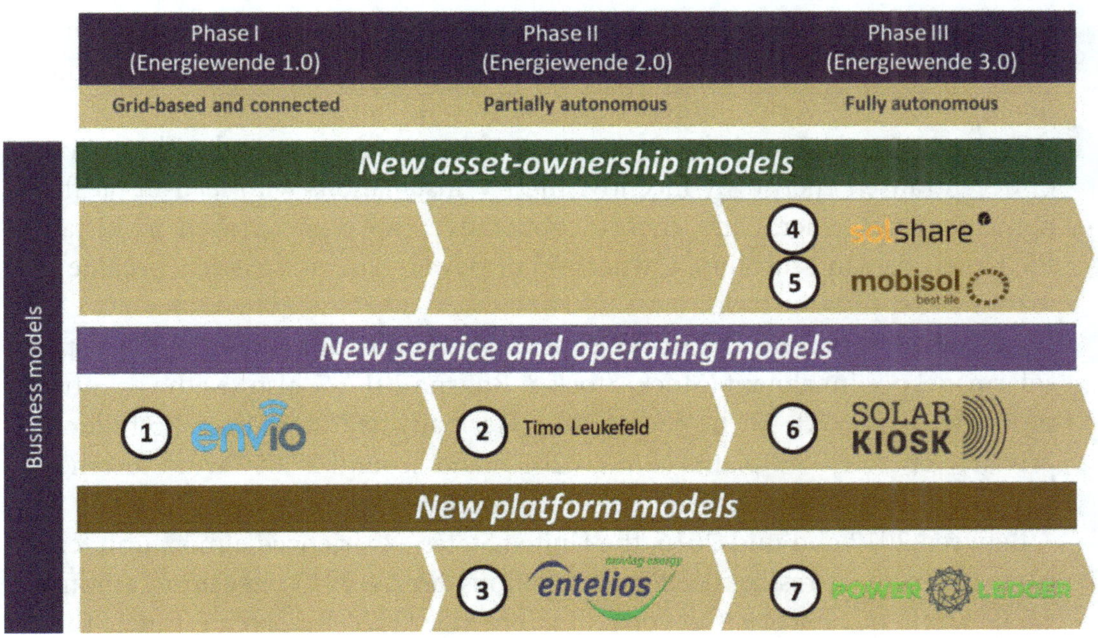

Figure 40: Business models and phases of the energy transformation

Source: Authors' contribution.

of electricity and natural gas supply – requires capital-intensive investments up front to establish pipelines and transmission lines; generation plants and transformers; metering devices at the final users' residences; data and billing centres; and many more technical features.

In the traditional configuration of the electricity supply industry, assets were owned by the state, state-owned enterprises and municipalities, or by regulated private entities. Following the rapid expansion of energy demand in the 1980s and 1990s – especially in developing countries – new entrants from the private sector were able to build, own, and operate – or to build, own, and transfer – assets under long-term contracts with utilities. Simultaneously, many countries in the Western hemisphere started liberalising their power and gas sectors and opened parts of the value chain to competition, especially generation and retail.

Since liberalisation, the public ownership of assets has decreased. The large-scale privatisation of distribution and generation first occurred in Latin America and the Caribbean – starting with Chile under Pinochet in the 1980s – followed by Eastern Europe and Central Asia after the fall of Communism in the 1990s and 2000s.

Privatisation typically implied transferring ownership from a public enterprise to a multinational investor or utility that would take over the assets. For example, European energy incumbents such as Finnish utility Fortum, German E.ON, or Italian Enel entered the Russian market when the state-owned assets were sold. However, with liberalisation, a substantial number of new players entered the electricity market. In particular, newly created wholesale markets encouraged traders to start dealing with electricity and natural gas as commodities; insurance and hedging instruments were offered by players from the financial services world. In these cases, they would not *own* the assets.

The most fundamental change in terms of ownership occurred with the rise of decentralised energy. By incentivising PV installations with feed-in tariffs, homeowners, farmers, and energy cooperatives were encouraged to install PV panels or wind turbines. Ownership of power-generating assets became a mass-market phenomenon.

Australia has become the world leader in rooftop solar, with more than one solar panel per inhabitant (Stock, Stock & Bourne 2017: 7, also see the country profile in this book). At the end of 2016, Germany had more than 1.5 million privately operated PV plants (BMWi 2017) and around 27,000 wind turbines (BWE 2017), all of which were overwhelmingly owned by private entities.

Between 2012 and mid-2016, the United States experienced almost uninterrupted growth in residential PV installations (Perea 2017), reaching around 2.3 GW in 2017. Forecasts of consulting practice GTM Research estimate this capacity will grow to more than 4 GW in 2022 (ibid.).

Asset ownership can also materialise in energy associations or energy cooperatives. By contributing a certain amount of money, citizens who do not have the possibility to set up renewable-energy facilities on their land or rooftops can donate a certain amount of money and participate in an association. In Germany, the number of energy associations rose from 66 in 2001 to more than 850 in 2017, and total membership is around 180,000 and growing (Bundesgeschäftsstelle Energiegenossenschaften 2017). A survey by the German Cooperative and Raiffeisen Confederation DGRV (2015) reveals that more than a fifth of the members pay €1000 or less to become part of the association, and around a third pay between €1000 and €3000. Hence, even for people with smaller budgets, it is easy to be a co-owner of a PV plant or a wind turbine.

Energy associations typically have a geographical proximity to their assets. They are often part of a local initiative, with residents, regional banks, and even municipalities sometimes serving as the driving forces. Meanwhile, digitalisation has made it possible to detach ownership from location. Over the last decade, crowdsourcing platforms have mushroomed, often to raise money for cash-strapped entrepreneurs. The most popular platform was Kickstarter, through which the creators of the Fairphone as well as Elon Musk and his electric car company, Tesla, raised money for their ventures. Crowdfunding typically follows a simple principle: anyone in the world is invited to invest money in a certain project – often as an upfront investment – and receive a product or service during the later stages of the venture. In the energy sector, crowdfunding platforms such as 'crowdener.gy' or 'econeers' raise money for specific projects, most often in the field of renewable energies.

Asset ownership, hence, has become a globally dispersed phenomenon. As opposed to donations to charities or non-governmental organisations, in crowdfunding the intermediary is digitalised – donors and receivers directly interact with each other.

In the near future, though, asset ownership may advance to the sphere of cryptocurrencies and decentralised ledger protocols such as blockchain. Owning real assets with a virtual currency sounds like a logical inconsistency. However, the value of a cryptocurrency is based on a consensus of the value of the currency among those who have invested in it. It is a system of faith and speculation, similar to most market-based assets.

The buzz term in this dimension is 'initial coin offering' (ICO). In an ICO, the issuer offers so-called tokens (Orcutt 2017). A token can represent a certain amount of a virtual currency, but it can also be cloud storage space, as in the start-up Filecoin, or it can correspond more closely to traditional stocks of the issuing firm, as in decentralised autonomous organisations (or DAO, see Burger et al. 2016 for a discussion). In the energy industry, Australian start-up Power Ledger raised US$24 million in October 2017. Around 15,000 supporters invested in the ICO. Power Ledger's tokens are called POWRs and are convertible into Sparkz, which is the virtual currency for Power Ledger's users to trade electricity among themselves (St. John 2017). Another application suggested by Power Ledger is 'autonomous asset management'. According to information issued for Power Ledger's token generation event in October 2017, this application provides a platform for shared ownership of renewable-energy assets as well as for trading renewable-asset ownership (Power Ledger 2017).

Of course, ICOs in the energy sector are still niche applications of a niche technology. In addition, public authorities such as the US Securities and Exchange Commission and, in the case of Power Ledger, the Australian Securities and Investments Commission (ASIC) are exploring how to deal with ICOs from a regulatory perspective. For example, ASIC interprets ICOs as a type of 'Managed Investment Scheme and therefore subject to the Australian Corporations Act' (Thomsen 2017). With multiple applications foreseeable, ICOs may become the virtual mirror image of the dispersed ownership of decentralised energy assets.

Sebastian Groh's insights from Bangladesh reveal that decentralised asset ownership can be particularly advantageous in the context of developing countries, where a connection to the central grid represents a certain elevation in social status, but may also coincide with a lower quality of power supply because of multiple interruptions. He observes that swarm energy based on micro-grids proves to be more resilient and reliable than the central grid. Via their mobile phones, those who own their own PV systems can use the money they earn from selling electricity in real time. However, the success of these systems depends on the regulatory framework and the willingness of governments to consider decentralised energy as an acceptable alternative to the central grid and provide financing schemes for rural residents.

Klara Lindner's contribution on Mobisol's success story shows that start-ups operating in this difficult environment of rural, decentralised electrification need to be in control of the whole value chain – including devices such as TVs and mobile phone chargers, which they sell in conjunction with battery and rooftop solar – to ensure quality and reliability.

New service and operating models

Long before the liberalisation and privatisation of the electricity sector, companies specialised in service operations, optimisation, and the maintenance of energy installations, in particular in the area of building efficiency. The most prominent example is energy performance contracting (see also Burger & Weinmann 2013 for a discussion). Since the 1970s, energy service companies have assisted utilities as sub-contractors to increase energy efficiency, typically based on energy audits. A new industry emerged that was able to deliver turnkey projects for large industrial and institutional customers (IFC 2007; Li, Qiu & Wang 2014). According to the National Association of Energy Service Companies of the United States, the industry experienced a period of stagnation after the collapse of ENRON in the mid-2000s (IFC 2007: 9), but recovered soon afterwards. Based on a market survey by US company Navigant, the European Commission's Joint Research Council publishes estimates of the European market to be around €2.4 billion, with moderate growth of 1.7 per cent per year by 2024 (Boza-Kiss, Bertoldi & Economidou 2017).

Business models in energy performance contracting may include the financing or leasing of assets, but typically the owner of the dwelling or industrial facility also owns the assets, which the contractor has installed, after the contracting period has ended.

As the increasing deployment of renewable energies leads to a more inelastic primary energy supply, it also becomes more attractive for companies, households, and utilities to increase their individual elasticity of demand and exploit financial opportunities related to fluctuations in wholesale market prices. Business models based on efficiency services and the optimisation of the operation of energy assets – including the equipment of dwellings, such as lighting and windows – require in-depth knowledge of the complex interplay of all the energy-related components of each object. With the use of practically unlimited computing power and artificial intelligence (AI) algorithms – also called machine learning – the barriers to entry in this field have been significantly reduced, and new players can more quickly and easily access the market than before. For example, MeteoViva is a start-up that uses local weather forecasts to optimise heating and cooling. The co-founder of Envio Systems has presented his company's business model in this book. He and his colleagues have developed a low-cost solution – compared to incumbent multinationals such as Johnson Controls, Schneider Electric, and Siemens – for retrofitting existing commercial and industrial buildings.

Not only new entrants, but also European energy utilities are moving towards service models. As in many other industries that are also moving from a 'technology push' to a 'market pull'. Customer-centricity has become a core element of their new strategies. For the electricity supply industry, this step is particularly challenging. Before liberalisation, residential customers were represented as standardised load profiles, anonymous numbers that could be easily aggregated, with minimal interaction occurring between the utility and user except for billing and meter readings. As regional monopolists, utilities did not have to take the differing preference sets of customer segments into account. With liberalisation, though, residential consumers started to become a relevant, non-negligible factor, since they could choose between different suppliers. Still, most utilities concentrated on defending their incumbent positions by offering differentiated tariff schemes adapted to each household's preferences. Most importantly, the cost and origin of electricity supply became the differentiating factors. Utilities realised that some customer segments were willing to pay a premium for climate-friendly primary energy sources, whereas other segments would choose primarily based on a comparison of costs per kilowatt hour.

The situation has fundamentally changed now. New entrants from other industries, in particular companies operating in information and communication technologies, such as Google, Apple, and telephony operators, have discovered electricity supply as a service they can provide in a package together with home automation, entertainment, as well as safety and security features. Building efficiency becomes part of a larger bundled service, for example with devices such as the intelligent thermostats provided by Nest and Tado, the manufacturers of heating, ventilation, and air-conditioning control systems equipped with – and linked to – machine learning algorithms, and aesthetically appealing as lifestyle gadgets. In early 2014, US company Google acquired Nest for US$3.2 billion (Heisler 2016). Even though the acquisition might not have resulted in returns that would justify the high price tag, according to Heisler, Google's intention was to become a holistic service provider with access to energy consumption data and, more generally, the Smart Home.

Most of the pilot projects and large-scale field trials, which provide more complex services than the previous standard uniform tariffs, focus on differentiated electricity prices (day – night, weekdays – weekend, summer holidays – regular working weeks, or even more fine-grained differentiations of retail tariffs according to actual wholesale market prices). However, they have not yielded satisfactory results in terms of energy savings and residential demand shifts. Even the European Commission admits that residential consumers are only shifting 3 per cent of their demand when tariff schemes include a financial incentive to shift demand (European Commission 2018). Hence, business models that focus solely on shifting and optimising residential demand suffer from the fact that this type of demand is more inelastic than anticipated. The differences in tariffs simply do not provide a sufficiently rewarding incentive for consumers to change their habits when washing clothes, boiling water for tea, or baking a cake. Uniform tariffs are comparable to an insurance premium that residential consumers are willing to pay in order not to bother about the real (i.e. wholesale market) price of power. One has to bear in mind, though, these are very early days in terms of automation of household appliances, and it may get easier with new generations of white goods to access value at that granular level of residential, domestic consumption.

By contrast, service models that include other factors of convenience, such as making the home safer or allowing for assisted living, are more likely to attract the attention of customers. Utilities have realised that their expertise in these fields is limited. Even during liberalisation, their willingness to enter alliances was limited. They instead set up their proprietary business units, such as in trading, but now utilities have start establishing alliances. For example, German utility EnBW has founded a company called Qivicon, together with partners from telecommunications as well as white goods and Smart Home device manufacturers.

Service and operating models become attractive for a few reasons:

- Cash-strapped utilities may no longer have the financial leverage and shy away from investing in capital-intensive infrastructure, such as large-scale power plants. In addition, they fear stranded assets and the risks associated with an uncertain regulatory and market environment.

- By contrast, they do have the expertise for providing services, at least in the field of energy, within their existing workforce. Often that type of human capital is dispersed across different business units, though, and has to become reorganised to provide a single point of contact for customers.

The less capital-intensive nature of these service-oriented business models makes it easier for new entrants and start-ups to enter the sector, too. The disruptive changes of digitalisation lower barriers to entry because hardware becomes less important than software, AI algorithms, and the customer interface.

Envio Systems is one example of a company that has utilised the hardware-as-a-service model, much like a marketing tool to attract cash-constrained customers to enter a service contract with them. Even though their payback period is substantially shorter than their high-end competitors, such as Siemens and Schneider Electric, the case of Envio Systems shows that the market for retrofitting existing buildings is still challenging.

Timo Leukefeld confirms this view. By contrast, his strategy is building and renting out new single-family and multi-story houses that are equipped with PV panels, solar thermal installations, a stationary battery, and a large water reservoir to store the heat. In this case, the configuration of the dwellings can be optimised according to the energy-efficiency standards. His practice cooperates with real estate developers, banks, and utilities. It is highly successful because tenants pay only slightly higher rents than in conventional houses. He believes in the 'flat rate society', where energy is part of a larger convenience package.

In the developing world, service and operating models are prospering, too. Solarkiosk has chosen a franchising approach of renting out its kiosks, because otherwise they would be too expensive for the shopkeepers in the countries they are targeting. It also leaves a maximum amount of autonomy to the operators of the kiosks – namely which products they want to sell and which services they offer – because the founders of the start-up believe that local residents would know best about the demand structure. Solarkiosk also shows that cooperations can be highly beneficial, as is the case with having Coca-Cola as a financial partner.

New platform models

Even before the liberalisation started, the electricity (and to some extent also natural gas) sector had some characteristics of a platform model: An entity – in most cases the grid operator, the utility, or a regional transmission operator – was in charge of coordinating multiple generation units and power plants, which were sometimes owned by private operators. As opposed to grid operations, which are a critical part of the infrastructure with elements of a natural monopoly, liberalisation in the generation part of the value chain added wholesale markets as trading platforms. Organisations such as the European Energy Exchange in Leipzig, with its subsidiaries and partners all across Europe, are providers of the IT infrastructure, but neither own assets nor operate them.

In Phase I of the energy transformation, they are the relevant references for buyers and sellers of energy, either directly or indirectly via brokers. When a country enters Phase II of the transformation, regulation often changes from fixed instruments such as feed-in tariffs to market-based mechanisms to compensate owners of renewable-energy assets. They have to develop their own strategies for how to earn money on the market, or they can shift that responsibility to aggregators. These companies typically offer a package of services to the owners that is centred around the timing and duration of operating the assets. They combine individual assets into virtual power plants. In Germany, Next Kraftwerke is the largest aggregator, combining a total of more than 6800 generation units, most of them renewable energies, with a capacity of almost 6 GW under its digital umbrella (as of February 2019).

This service is not limited to generation units. It can also integrate Demand Response for peak-shaving or ancillary grid services such as balancing energy, which become increasingly important when the share of renewable-energy intake rises above certain thresholds. Oliver Stahl, the founder and former CEO of the Demand Response solution provider Entelios, was interviewed for this book. His company not only aggregates and coordinates participants in the pool, but also offers this service to utilities that may not have the relevant expertise. In the interview, Stahl describes three major hurdles for Demand Response. The first is to make potential customers aware of the financial attractiveness of temporarily reducing their demand and gaining their confidence. Entelios picked one customer per industry sector as a role model in order to convince other players from the same sector. Second, not all industries are equally well-suited for Demand Response. In particular, the chemicals industry showed some hesitation because many processes are highly sensitive to fluctuations in energy supply. Third, becoming an integrated segment of a utility did not work because the representatives of the sales department – the most likely business unit to internally host a Demand Response provider – had an antagonistic incentive system and working culture.

In Europe, one of the largest companies offering Demand Response to commercial and industrial consumers is the Belgian start-up Restore, which targets the primary reserve and frequency control markets and operates in all ancillary services and capacity markets in Europe.

Some countries that have started moving from Phase II to Phase III of the transformation offer possibilities to establish peer-to-peer trading platforms. Australian start-up Power Ledger is one of them. Individuals can trade energy without interference from a utility as the intermediary. Depending on the regulatory context, these options are easier to implement in separate micro-grids, which are semi-detached from the central grid, because there is a higher degree of administrative autonomy.

Platform models have become very popular in many parts of the business world. Hospitality, accommodation, and transport services are offered that connect private individuals with other private individuals. Once energy services become largely decentralised, platforms may not only allow for trading, but also for more complex interactions. Australian start-up Power Ledger envisages smart demand and supply management, whole-market settlements, carbon trading, and even certain services in the management of transmission networks, such as network load-balancing, as future applications of its platform.

3.10.2 Six core competencies for corporate decision makers

How can corporate and political decision makers optimise their actions towards the transformation of the energy sector? The interviews with founders and entrepreneurs in the field of decentralised energy systems yielded important insights.

Digitalisation

The core competency that all industries are currently establishing is expertise in digitalisation – a theme that is present in all interviews – be it with artificial intelligence in the case of Envio Systems, the remote operating centres of Entelios, Mobisol, and SOLshare, or the blockchain-based application platform of Power Ledger. Beyond the capability of navigating in the digital sphere, the authors have identified five core competencies that will be decisive in tomorrow's decentralised energy markets.

Customer centricity

Customer centricity is an overarching, recurring theme in all of the interviews, which comes as no surprise, given that customers are coming into the focus of companies across almost all sectors.

Some of the start-ups presented in this chapter have customer-centricity within their corporate DNA. For example, demand-side management can only succeed if clients are treated individually in their energy-consumption and energy-savings patterns. Similarly, Envio Systems is developing a digital clone of each dwelling that is equipped with its Cubes, because retrofitting existing building stock requires a high degree of customisation – as opposed to building efficiency being implemented, for example, in greenfield real estate developments, where certain technological features can be replicated across all objects. Solarkiosk leaves it up to the operators of its kiosks as to which products and services they want to sell, since these people have a better understanding of the local context and the needs of future clients. Their approach combines the two worlds of charity projects with business ventures into a social enterprise. As Lars Krückeberg, co-founder of Solarkiosk, comments with regards to the context of developing countries, 'The combination of the two worlds is the only implementable option that allows for scalability'.

Other start-ups have developed standardised solutions but use a customer-centric approach in refining their technologies and business models. For example, Klara Lindner reports that Mobisol developed a plug and play kit in co-creation sessions with real customers. Later they realised that their customers were not willing to do the installations by themselves, so they adapted their business model to accommodate that wish by hiring local village technicians, who would be in charge of the installations.

The core competency required in all these areas is not customer-centricity *per se*, but rather finding the *balance* between listening to users while ensuring a high degree of standardisation. Customer-centricity comes at a price, and the core competency required is to drive costs down by developing new forms of mass customisation.

Financing and enabling of asset ownership

Start-ups have developed diverse strategies for how they can help prosumers to finance decentralised generation assets. Especially in developing countries, the major hurdle of a large upfront investment has been removed; in rural settings, residential owners of rooftop solar systems can generate additional revenues – for example by charging their neighbours' mobile phones or selling electricity on the micro-grid – to repay their debts, as the interviews with SOLshare and Solarkiosk and the contribution about Mobisol show. Start-ups use advanced transaction systems that are adapted to local markets, with the additional advantage of full transparency of money flows via apps on their mobile phones. For instance, SOLshare co-operates with bKash, currently the largest mobile money provider.

Financial competencies stretch into the sphere of cryptocurrencies. Australian start-up Power Ledger does not rely on Bitcoin, the most popular and widely known cryptocurrency, but enables transactions with a virtual currency called Spark, which consumers can use to purchase electricity or as producers to sell it. Power Ledger also spearheads the financing revolution in terms of crowdfunding and ICOs. In October 2017,Power Ledger raised more than €20 million through its token generation event. More than 15,000 buyers took part in Australia's first ICO.

Technology leads and product innovation

Despite the fact that the power supply industry is moving from product to service orientation, all start-ups have developed intellectual property in terms of product innovations. Envio Systems relies on its Cube, which is equipped with sensors like a CO2 sensor, to detect whether any person is actually present, and then uses proprietary artificial intelligence to optimise the consumption patterns of its clients. Mobisol works with its own direct current (DC) system, which is complemented by the household appliances of partner manufacturers running on DC. SOLshare produces its SOLbox and SOLcontrol devices directly in its workshops in Bangladesh. Solarkiosk developed and tested the containers that would later be transformed into kiosks under extreme climatic conditions in laboratories of the Beuth School of Applied Sciences in Berlin before the first kiosks were shipped to Africa. Power Ledger has built a proprietary IT architecture based on blockchain transaction protocols. Entelios also uses its own software to coordinate automatically the loads of its clients.

Technological advances and innovations give start-ups the leading edge and competitive differentiation. If companies rely on a pure service model, they can easily be crowded out by larger and financially stronger rivals.

Partnerships and bundled services

In a complex and highly dynamic market environment, no single company is able to provide all the elements of its value proposition by itself. Especially in the regulated world of utilities, which were the sole providers of energy, partnerships were not necessary because tariffs were a product of negotiations between suppliers and regulatory agencies or the government. Liberalisation has opened up the market to a range of combinations – packaged offers that may include entertainment, security, and individual transport in addition to core energy services. Consumers may enter, as Timo Leukefeld calls it, 'the flat rate society', where renting a house includes all turnkey solutions of convenience, plus free access to electric mobility. Leukefeld cooperates with real estate developers, banks, and utilities to build his dwellings. Energy is the trigger and remains an important element of his value proposition, but it is complemented by other products and services.

Similarly, Solarkiosk has teamed up with Coca-Cola, Ericsson, and Total. Envio Systems is seeking partners in major property management companies and with an elevator manufacturer. SOLshare has strong ties to Grameen Shakti, with whom it received the US$1 million UN DESA Powering the Future We Want prize. The municipal utility of Munich became the first industry partner of Entelios and already signed a contract even before its founder, Oliver Stahl, had approached venture capitalists for additional financing. In its international expansion, Power Ledger has deals, for example, with BCPG, one of the largest solar IPPs in Thailand; Tech Mahindra in India; and the Liechtenstein Institute for Strategic Development, which will become the first application host to offer Power Ledger's blockchain-based peer-to-peer energy-trading platform in Europe.

With the increasing convergence of the energy and transport sectors, digitalisation affecting all aspects of our lives, and US tech giants entering energy markets, executives face no other option than to enter partnerships and alliances if they want to survive in the marketplace.

Platforms/ecosystems

Digitalisation allows multiple players to enter markets and match supply and demand. As companies such as Uber, Airbnb, and eBay have demonstrated in other sectors, ownership of physical assets may not be necessary to succeed in the marketplace. The value proposition is derived instead from the coordination of providers and seekers of certain services. Sometimes these markets do not exist and have to be established: Entelios triggered the market for Demand Response in Germany, as much as EnerNOC contributed to similar developments in the United States. Start-ups and cash-constrained utilities may seize the opportunity to provide the IT architecture to connect sellers and buyers, often with technologies that circumvent conventional routes.

Australian start-up Power Ledger has demonstrated with its token generation event that platform applications will change how the electricity supply industry functions, for example: peer-to-peer trading; smart demand and supply management with remuneration and payment settlements; management of consumer exposure to the risk of non-supply; collection of big data; Smart Contracts for carbon traders; management of transmission networks; network load-balancing; and power ports that allow electric vehicles to become mobile storage discharge facilities.

Many ecosystems do not yet exist, but those companies and governments that foresee their benefits and start with the implementation may be the winners in the energy world of the future.

3.10.3 A world of entrepreneurial activity

The business models of the seven start-ups that have been presented in this chapter are of course only a fragment of the wide spectrum of entrepreneurial activity that has emerged across the globe. Start-ups in all continents seize the opportunity of decentralisation to launch their ventures – some of them focusing on new business models with existing technologies, others developing new generation technologies or platforms. Innovation has moved from research labs and R&D departments of manufacturers and utilities to a generation of young ventures – with digitalisation as the major driver for reducing barriers to entry.

As outlined in Section 3.1 of this chapter, the global energy transformation can be differentiated according to three distinct phases, which may vary and overlap across countries and regions. The next chapter summarises the insights from both regulatory conditions and business models within the respective phase.

3.10.4 References

Boza-Kiss, B., Bertoldi, P. & Economidou, M. (2017), Energy service companies in the EU: status review and recommendations for further market development with a focus on energy performance contracting. EUR – Scientific and Technical Research Reports. Ispra: Joint Research Centre (JRC).

Bundesgeschäftsstelle Energiegenossenschaften. (2017), Die Genossenschaften in Deutschland. Retrieved November 24, 2017, from https://www.genossenschaften.de/bundesgesch-ftsstelle-energiegenossenschaften

Burger, C. & Weinmann, J. (2013), The decentralized energy revolution – business strategies for a new paradigm. Basingstoke, UK: Palgrave Macmillan.

Burger, C., Kuhlmann, A., Richard, P. & Weinmann, J. (2016), *Blockchain in the energy transition. A survey among decision-makers in the German energy industry. Berlin: German Energy Agency dena/ESMT*. Retrieved from https://shop.dena.de/sortiment/detail/produkt/studie-blockchain-in-der-energiewende/

BWE. (2017), *Anzahl der Windenergieanlagen in Deutschland*. Retrieved November 20, 2017, from https://www.wind-energie.de/infocenter/statistiken/deutschland/windenergieanlagen-deutschland

DGRV. (2015), *Energiegenossenschaften – Ergebnisse der DGRV-Jahresumfrage* (zum 31.12.2014). Retrieved February 14, 2019, from https://www.dgrv.de/de/news/news-2015.07.16-1.html#

European Commission. (2018), *Smart grids and meters*. Retrieved March 7, 2018, from https://ec.europa.eu/energy/en/topics/markets-and-consumers/smart-grids-and-meters

Gassmann, O., Frankenberger, K. & Csik, M. (2014), The business model navigator: 55 models that will revolutionise your business. London: Pearson/Financial Times.

Hamwi, M. & Lizarralde, I. (2017), A review of business models towards service-oriented electricity systems. *Procedia CIRP, 64*, 109–114.

Heisler, Y. (2016), *Google's Nest acquisition was more disastrous than we thought. BGR*. Retrieved November 30, 2017, from http://bgr.com/2016/06/06/googles-nest-acquisition-was-more-disastrous-than-we-thought/

IFC. (2007), *Introduction to energy performance contracting, National Association of Energy Services Companies*. Retrieved February 11, 2019, fromhttps://www.energystar.gov/ia/partners/spp_res/Introduction_to_Performance_Contracting.pdf

Li, Y., Qiu, Y. & Wang, Y.D. (2014), Explaining the contract terms of energy performance contracting in China: The importance of effective financing. *Energy Economics, 45*(C), 401–411.

Orcutt, M. (2017), What the hell is an initial coin offering? Cambridge, MA: MIT Technology Review.

Perea, A. (2017), *Q3 2017 Update: the state of distributed solar. GTMresearch*. Retrieved February 11, 2019, from http://www.ncsl.org/Portals/1/Documents/energy/webinar_A_Perea_9_2017_31661.pdf

Power Ledger. (2017), *Power Ledger token generation event*. Retrieved November 27, 2017, from https://powerledger.io/

St. John, J. (2017), *Blockchain energy trading startup Power Ledger raises $17M in cryptocurrency 'ICO' GTM Research*. Retrieved February 11, 2019, from https://www.greentechmedia.com/articles/read/power-ledger-blockchain-energy-trading-startup-raises-17-cryptocurrency#gs.u0lJtXV5

Stock, P., Stock, A. & Bourne, G. (2017), *State of solar 2016: globally and in Australia, Climate Council.* Retrieved February 11, 2019, from https://www.climatecouncil.org.au/resources/solar-report/

Thomsen, S. (2017), Australia's first initial coin offering raises $34 million. *Business Insider Australia.* Retrieved February 11, 2019, from https://www.businessinsider.com.au/australias-first-initial-coin-offering-has-raised-34-million-2017-10

CHAPTER 4

The three phases of the energy transformation – combining governance and business model innovation

This chapter merges the findings of the two preceding chapters – on the importance of governance and the simultaneous development of business models – in moving from the conventional, fossil fuel-based energy system to one that is decentralised, renewable and based on energy efficiency. It highlights characteristics of three distinct phases during that transition and concludes that it is vital to align governance (policies, network rules, market rules, institutions etc.) with flexible system operation to achieve a smooth transition. In the context of developing countries with incomplete grid infrastructure, regulatory incentives and business initiatives may lead to a leapfrogging effect from Phase I to Phase III, though.

4.1 Three phases of the transformation

The transformation towards a decentralised energy system entails both regulatory incentives as well as entrepreneurial initiatives. There are three broad phases in moving from the conventional, fossil fuel-based energy system to one that is decentralised, renewable and based on energy efficiency.

In technical terms, Phase I can roughly be associated with a niche deployment of decentralised renewable energies; in Phase II, their contribution rises to become a major player in the supply portfolio, enabled by governance which provides value for the necessary system flexibility requirements; and Phase III is characterised by decentralised renewable energies as the dominant player within a flexibly operated system.

The following table describes the main features of each phase.

How to cite this book chapter:
Burger, C., Froggatt, A., Mitchell, C. and Weinmann, J. 2020. *Decentralised Energy — a Global Game Changer.* Pp. 261–275. London: Ubiquity Press. DOI: https://doi.org/10.5334/bcf.t.

Table 5: Overview of the three phases of energy transformation.

	Phase I	Phase II	Phase III
	Grid-based and centralised system with decentralised renewables as a niche phenomenon (contribution to total power generation less than 10 per cent)	*Decentralised renewables growing in importance (contribution to total power generation up to 40 per cent)*	*Decentralised renewables as dominant player with fully autonomous solutions not connected to a central grid*
Governance	**Centralised system regulation** • Promoting renewables • Promoting local and national industry • Promoting lead markets	**Performance based regulation** • Evaluating decentralised renewable generation versus network costs • Flexibility and energy efficiency incentives • Integrating customers • Integrating electricity, heat and mobility (convergence)	**Consumer focused ambition driven regulation** • Users can choose the level and method of security of their supply • Co-existence of central and decentralised systems and regulations
Business models	• New asset ownership models: from central to crowdfunding		
	• New service and operating models: from bundled to autonomous operations		
	• New platform models: from aggregators to open platforms		
Core competencies	• Financing and enabling asset ownership • Technology leads and product innovation	• Customer centricity and meaningful consent • Partnerships and bundled services • Technology leads and product innovation	• Financing and enabling asset ownership (ICO) • Technology leads and product innovation • Platforms and ecosystems
Risks	• Low risk	• Risk of stranded investments in fossil/nuclear assets	• Risk of stranded investments in transmission grids

Source: Authors' contribution.

4.2 Phase I (Energiewende 1.0): grid-based and connected energy system with decentralised renewables as a niche (<10 per cent)

4.2.1 Governance of Phase I: centralised system regulation promoting renewables, local industry, and lead markets

Most countries in the world have ratified targets for the rollout of renewable energies (Ren21 2017a). The encompassing objective typically is the move towards less carbon-intensive energy supply, thus contributing to the reduction of greenhouse gases.

Governance in Phase I is driven by incentivising the deployment of renewables. In some countries, the introduction of regulatory incentives to promote renewable energies was accompanied by objectives related to industrial policy, in the attempt to create lead markets for solar cell manufacturers or wind turbine producers. For example, the Danish wind turbine industry and the German photovoltaic manufacturers greatly benefitted from direct and indirect subsidies and support schemes governments imposed to promote these technologies (Lipp 2007; Nicolini & Tavoni 2017; Strunz, Gawel & Lehmann 2016).

Some of these joint renewable and industrial policies were successful, for example the development of Danish wind turbine producer Vestas on the back of policies to support domestic wind energy. Vestas is still among the world's leading wind turbine manufacturers, providing a role model of where public and corporate interests have been aligned.

Other incentives such as those for solar photovoltaics in Germany led to short boom for some German photovoltaic manufacturers. In the longer term, those manufacturers, for example SolarWorld (DW 2017), could not compete with companies from countries with much lower unit production costs, in particular China. The ambition of the German government to create an East German 'Solar Valley' in the state of Brandenburg did not yield the expected results. Most of the companies either went into bankruptcy when the incentives were reduced, or were acquired by Asian competitors (Fuchs 2015). However, it has also led to the development of 'second' generation photovoltaic knowledge in Germany, as well as the development of inverters and fittings – which make up 40 per cent of the cost of the panels (Wirth 2018). In the longer term, higher shipping costs and long delivery times may improve the competitiveness of German and European manufacturers vis-à-vis their Asian competitors (ibid.).

Other countries, such as the Netherlands, decided to follow a wait-and-see strategy and initiated incentive schemes only after the costs per kilowatt hour were substantially reduced.

The advantage of that strategy is that it avoided any boom-and-bust of Dutch manufacturing industries and saved direct subsidies, that is, Dutch taxpayers' money. The benefits have to be counterbalanced, though, with potentially negative consequences for countries like the Netherlands in terms of innovation capabilities, because of the limited involvement and experience with new energy technologies and system operation.

In Phase I, energy governance encompasses direct policies primarily in support of new capacity, but also other issues related to network access rules, market rules and design, retail competition rules, and so on. The most common policy incentives in support of renewable energy development are feed-in tariffs (FITs), net metering, and renewable portfolio standards. All instruments have their pros and cons (SSREN 2012):

- ***Feed-in tariffs*** (FITs) are based on a fixed remuneration per kilowatt hour fed into the grid, typically over a time horizon of ten to twenty years, and can be considered a long-term purchase agreement. FITs typically guarantee a fixed price per kWh, an offtake contract, and priority access to the network and electricity market. This means that risk is substantially reduced for investors, and it is possible to borrow money on the contract. There are multiple examples of FITs in different countries, including Australia, China, Denmark, and the United States, with Germany generally seen as the most successful case – see country Section 2.9.

  FITs tend to be technology specific, which means the remuneration given depends on the technology and its investment costs. With this instrument, governments can steer their support towards individual renewable sources which would not be competitive in market-based approaches, for examples auctions. Historically, solar power has benefitted from FITs.

  In Phase I of the energy transformation, feed-in tariffs provide market certainty for investors (Cox & Esterly 2016), but they may become too expensive in later stages and be replaced by more market-based instruments: 'Policymakers have recently been moving towards designing FITs as a premium in addition to the current market rate for electricity, known as feed-in premiums' (ibid.).
- Another Phase I type policy in support of scaling technologies is ***Net Energy Metering*** (NEM), also called Net Metering. It allows residential owners of photovoltaic panels to sell their surplus electricity to the local utility at retail rates – the meter in effect turns back for every kWh that is injected from home into the grid. In return, households benefit from a net reduction of their utility expenses. NEM originates from the United States, and has been a popular instrument there. As of November 2017, 38 US states, Washington DC, and four territories offer net metering incentives (NCSL 2017). Especially in states in the south-western part of the United States with high solar radiation or in states with high electricity tariffs, such as in the New England states in the north-eastern part of the country, net metering can significantly lower the electricity bill of prosumers.
- The third major instrument in Phase I are ***Renewable Portfolio Standards*** (RPS), also called Renewable Energy Standards – again, usually found

in the United States, and often in conjunction with net energy metering. Renewable Portfolio Standards force utilities to reach a certain percentage of renewable energies in their portfolio within a given time frame. Depending on their design – for example, through tendering or auctions – they are often intended to encourage competition between external providers as well as between technologies to achieve the maximum reduction of greenhouse gas emissions at the lowest price.

When choosing policy incentives, policy makers have to consider their effect on promoting centralised or decentralised generation infrastructure, illustrated in Figure 40. The set-up of the instruments may have a fundamental impact on the infrastructure. With quotas, capacity payments, or renewable portfolio standards investor-owned utilities (IOUs) tend to build central infrastructure in line with their current business model and core competencies, whereas FITs, Net Metering, or Demand-Side Response (DSP) tend to attract decentralised infrastructure enabling a faster transformation towards Phase III.

During this first phase of development, the system is dominated by incumbent generators and suppliers, and grid operations remain centrally controlled by transmission and distribution system operators, as they have been for previous decades. The total contribution of power generation from new renewables hovers below critical thresholds. There is very limited impact on system operation, system costs, average electricity costs or on displacement of fossil fuels and nuclear power in the merit order. Utilities, system operators, and private companies are able to use the first experiences of intermittent sources within electricity systems to learn to develop and apply forecasting models and develop processes for information exchange, billing and accounting, and potential operational issues (Baumgartner 2017).

However, although the actual percentage of decentralised renewable electricity may not be high, it is also possible for countries to find the ownership of their generation transformed during this first phase – as with Denmark in the 1980s and 1990s (see also Section 2.8).

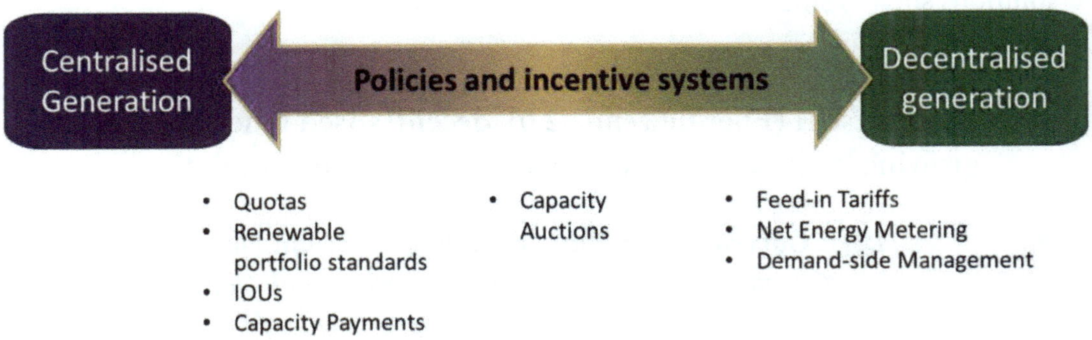

Figure 41: Policy incentives in centralised and decentralised systems.

Source: Authors' contribution.

4.2.2 Business models and core competencies of Phase I

In Phase I of the energy transformation, the traditional business model of power companies remains relatively unaffected by the deployment of renewables as their contribution remains low. Incumbent utilities continue to supply the vast majority of power, balancing against largely predictable demand fluctuations. The main investment uncertainties are about future demand and the price of fossil fuels.

Business models evolve around new offers in terms of financing, installation and operations of decentralised renewable energy assets. Private investors are attracted to participate in the energy market because of the low-risk purchase guarantees, available through instruments such as feed-in tariffs or net metering. Bioenergy villages and energy associations are founded, which rely on participatory decision-making and financial contributions from their members.

In this phase, it is not the volume of renewable energies that is transformative. For institutional investors, one of the most appealing attributes of renewables is that they are scalable. Cost reductions of renewable technologies have made even smaller-scale deployment economic, enabling small and medium-sized companies to enter into a market previously dominated by vertically integrated utilities, and small domestic or commercial level consumers to become prosumers. Consequently, decentralised generation, combined with energy efficiency and balancing of localised markets, creates the knowledge and human resource foundation for a broader decentralised supply industry.

Start-ups, utilities, and new entrants from other sectors, in particular information and communication technologies (ICT) and finance, gather expertise from the experience with first installations and pilot projects. In the first pioneering countries and states of the energy transformation, such as California, Denmark, and Germany, innovation has mainly focused on technologies and products, in particular the decline of unit costs of technologies such as solar cells and wind turbines, whereas in countries that have started later with programs to promote decentralised renewables, the momentum shifts from product to service innovation – with sophisticated financing models, integrated solutions such as smart homes, and a digital ecosystem.

4.3 Phase II (Energiewende 2.0): decentralised renewables growing in importance with partially autonomous solutions

4.3.1 Governance of Phase II: setting the ground for Energiewende 3.0 with performance-based regulation

When the system moves into Phase II, policy makers start modifying incentive systems, as the costs of renewable technologies fall and their deployment becomes more widespread. Simultaneously, government and regulators develop a deeper understanding of the key decisions they have to make if they want cost-effective, higher percentages of renewables (REN21 2017b). The early-mover countries are observed to abandon the initial subsidy schemes and establish more market-based mechanisms, such as auctioning, in parallel to FITs for smaller scale technologies. However, late-adopter countries are able to benefit from the impact of the first movers.

In this phase, the challenge for managing the governance framework is to ensure stable market conditions while creating a regulatory environment that encourages innovation and new market actors. The governance framework sets the policy direction and creates the regulatory framework to implement instruments that support and incentivise new actors or penalise uncompetitive practices, either via top-down or bottom-up processes.

In Phase II, governance has to lay the foundations for flexibility and energy efficiency, integrating consumers, developing mechanisms to deliver meaningful consent from people and society for the transformation, as well as electricity with heat and transport, and to evaluate network costs versus decentralised solutions. All these elements become crucial for a smooth transition to and as part of Phase III, where differing proportions of central to decentralised systems co-exist. Management of the process and expectations of the stakeholders involved becomes an important part to enable adaptation to new circumstances. Participatory approaches to integrate customers' requests and wishes into the public discourse are established, for example with regards to a higher density of wind turbines next to residential neighbourhoods, or the construction of new transmission lines (or the reinforcement of existing lines).

As the percentage of variable renewables increase, and if the system is operated in the same way as during Phase I, security of supply shifts to the attention of regulators and policy makers. The transmission and distribution grids have to be reinforced and expanded in certain areas. The distribution grid is affected, because conventional lines were laid out to unidirectionally satisfy residential demand. If entire neighbourhoods or even villages start producing their photovoltaic energy with decentralised units – and no adequate local storage solutions can temporarily absorb the power – supply may exceed local demand, and the electricity has to be transported to the next load centre. Large-scale wind power makes additional investments necessary at higher-voltage transmission levels, because – similar to photovoltaics in rural areas – wind power tends not to be produced in areas with high demand, such as urban agglomerations and their adjacent industrial sites.

Thus, increasing amounts of renewables requires new ways of regulation and operation of the system so that expensive additional transmission and distribution capacity can be kept to a minimum. The type of regulation and compensation mechanisms has a fundamental impact on how much additional transmission and distribution infrastructure capacity is required and the number of interventions needed by the grid operators, and this of course also affects the final cost of electricity to the customer (Shakoor, Davies & Strbac 2017).

As outlined in Section 2.9.4, the move to more flexible system operations is likely to reduce costs and ensure security of supply. It is not only a key requirement for Phase II but also the enabler of Phase III of the energy transformation. If the system becomes flexible very fast, Phase II may be substantially shortened and the transformation can occur in a non-disruptive way. Ensuring sufficient flexible resources and coordinating them efficiently is therefore a central aspect of the transformation from a fossil-based electricity system to a decentralised one.

Flexibility measures can include, for example, building short-term storage facilities, such as stationary batteries, improving industrial and residential demand-side response, reducing short-term and long-term energy consumption, and implementing local markets. Flexibility is closely linked to energy efficiency. Incentives to increase energy efficiency include both measures directed towards end users to enhance the efficiency of their buildings and appliances, as well as incentives to increase efficiency of energy system operations. The coordination of heat, electric mobility and decentralised power generation offers new possibilities to adapt demand to inelastic supply by intermittent renewables.

The least expensive solution to increased flexibility and efficiency during Phase II is demand-side response, though. As a platform model, it builds on the existing assets, exploits their flexibility potential, and brings down peak infrastructure needs and peak costs. The interview with Oliver Stahl, founder of German demand-response pioneer Entelios, in Section 3.5 of this book provides insights into the underlying mechanisms and business models.

Each of the policy interventions of Phase II will of course have to be counterbalanced with questions of privacy and data protection (see also Burger, Trbovich & Weinmann 2018). Until now, society has collectively embraced the transfer of private information as a compensation for services it gets for free by providers such as Facebook, Google, What's App, or Amazon. Sociologists call this willingness to share certain aspects of one's life the 'Privacy Paradox' (Wittes & Liu 2015; van Zoonen 2016; Wittes & Kohse 2017). Understanding what this means for society, and hence energy, is still in its infancy.

While there is a logical progression from Phase I to Phase III, how long any country needs to stay in Phase II is not pre-determined, and may in future be shortening as new technologies become economically viable and widely available. As Chapter 2 has shown, a key determining factor will be the governance framework and the efficiency of its coordination that can accelerate or slow down the rate of new technology deployment.

4.3.2 Business models and core competencies of Phase II

In Phase II of the energy transformation, the deployment of decentralised renewables becomes non-negligible and increasingly affects system and market operations.

The larger intake of renewable sources typically reduces wholesale market prices. On the European Energy Exchange, the spot-market power price fell to €29 per Megawatt hour in 2016, down from more than €70 per Megawatt hour in 2008. This trend is caused by the low operating costs of renewables, which crowd out more expensive coal and gas-fired thermal plants on the dispatch curve in the conventional marginal cost electricity market design (Thalman 2015).

In the short term, low wholesale prices are beneficial for consumers. If prices fall below certain thresholds of viable commercial operations, though, utilities may start mothballing or even dismantling central power plants. From a public perspective some of this may have positive effects – for example, the closure of coal plants to curb greenhouse gas emissions. However, flexible natural gas plants may compensate for supply fluctuations in the move towards systems with a higher share of decentralised renewable energies. In this situation, closure of flexible gas plants may not be in the mid-term public interest. Consequently, policy makers may modify the regulatory framework by setting up capacity markets for targeted flexible generation. In addition, they have to establish appropriate compensation mechanisms for stranded nuclear or fossil assets of utilities.

For conventional electric utilities, Phase II of the energy transformation leads to fundamental changes in their business model. Declining revenues from wholesale markets and thermal plants turning into stranded assets leads to financial stress. The orientation towards customer centricity and mind shift from large scale to small scale, in particular individualised energy efficiency solutions, cannot be expected to occur over night. Moreover, utilities have to deal with dwindling market shares in the new system, as new entrants threaten their positions. Digitalisation and falling technology prices enable players from other sectors and even start-ups to step in and establish themselves offering integrated solutions in the utilities' core markets. Utilities become aware that they cannot compete with tech giants and ICT firms on their own – hence, they start entering partnerships and alliances with telecommunication companies, providers of smart goods or start-ups to provide complex service solutions, for example in the smart home or electric vehicle charging markets. Product innovation then becomes a joint effort, connecting the digital ecosystem with smart devices.

As already highlighted, the introduction of feed-in-tariffs, especially for solar and wind, has enabled individuals, energy associations and local communities to invest in renewable energy. This is now expanding with consumers showing active engagement in combined storage and solar systems. New business models are enabling installations without, or with less, government support. In Germany and Australia, individual households with batteries and solar PV are being offered free electricity when they are unable to generate their own supply, in return for their batteries being used to maintain grid frequency – the creation of virtual power plants (Energy Brainpool 2016; Griffith 2017).

Increasing automation and data processing capabilities are leading to the development of new apps and platforms. A wide range of companies emerges, which automatically aggregate consumers buying power (such a Labrador power in the United Kingdom) and develop opportunities for flexible demand (such as Tempus Energy in Australia). These companies require consent from consumers to enable them to facilitate the automatic switch between suppliers and to vary consumption. The success of these companies is determined both by the regulatory environment in which they operate, and the on-going trust of consumers. Data management and automation is likely to reduce the need for active engagement of consumers to reduce their energy consumption; however, it will not eradicate it. If there is a societal consensus to curb greenhouse gas emissions, conscious lifestyle changes and consumer behaviour shifts are also needed.

The major change in business models during Phase II is the focus on customers' needs. Energy is no longer a commodity to be sold via the meter, it rather becomes a unique combination of technologies adapted to each individual rooftop (with respect to PV installations), to the topography of the territory (with respect to wind turbines), or to the agricultural intake from local farmers for biomass co-generation technologies. Decentralised deployment of renewable energy technologies supports local businesses and technicians, thereby creating local value.

4.4 Phase III (Energiewende 3.0): decentralised renewables as dominant player with fully autonomous solutions

4.4.1 Governance of Phase III: consumer-focused, ambition-driven regulation

The third phase of the energy transformation, or 'Energiewende 3.0', is characterised by decentralised renewable energies becoming the major player in the supply structure. Phase III is yet to exist in any developed country. Some countries and regions, such as Australia, Denmark, Germany, Ireland, as well as some states in the United States, experience prolonged intervals in electricity supply when new renewable energies account for the main source of primary electricity. For example, in March 2018 Portugal's renewable energy production exceeded power demand and accounted for more than 100 per cent of mainland electricity consumption (Reuters 2018). However, so far, even countries with higher proportions of decentralised renewable generation have not moved to becoming highly flexible energy systems. This flexibility characteristic can be seen as a key determinator of moving from Phase II to Phase III. Governance – meaning market design, network rules and incentives, tariff policy, its coordination and so forth – has been changed by the time a country becomes a Phase III country.

In rural settings of countries in the developing world, such as India, key characteristics of Phase III already exist, albeit often on a very low level of the energy ladder, and with substantial financial disadvantages for local residents.

The transformation is likely to occur in most geographic contexts, but its dynamics may vary across countries and continents. Different phases might co-exist across regions of a single country, for example in urban versus rural settings. Some countries might opt for largely remaining within the Phase I configuration, for example, if energy generation is predominantly based on large, central hydropower dams, which often produce cheap and climate-neutral electricity. Especially in developing countries with low electrification ratios, companies and governments may be able to leverage lessons learned in a Phase I environment and leapfrog directly to Phase III with largely autonomous island systems.

In the first two phases of the energy transformation, the pace of deployment of renewables is to a large extent determined by the national governance framework, including the ability of local administrations to be involved.

As electrification occurs in new sectors, such as transport, and to improve the efficiency of electricity generation, the governance framework needs to enable the co-existence of different systems: central grid-based, autonomous decentral entities and regions, further integration of heat and e-mobility. As the energy system decentralises, the importance of a distribution level system coordinator increases, to not only manage technical operations in the interest of all stakeholders, but also to stimulate new markets and thus enable new entrants and innovators. The increasing coordination function of the distribution system operator, leads to their development as a platform provider or Distribution System Platform.

At the heart of the future system must be performance-based regulation, which not only ensures supply obligations to be met, but also wider social and environmental sector goals to be delivered. This means that reforming regulation will encompass how to deal with winners and losers – and this is more than creating an open and transparent decision-making process. Society as a whole has to enter a public discourse over price stability and security of supply, jobs and corporate interests, climate change and effects on the local environment, for example wind turbines or large PV fields, and how to deal with stranded assets, in particular obsolete long-haul transmission lines.

Regulatory instruments, such as the introduction of capacity payments discussed in Phase II, can give rise to supply stability, but may also, in some cases, support incumbents' fossil-fired power plants that were scheduled to be mothballed because of environmental reasons. With intermittent solar and wind intake, there will be a resource abundance at certain times, and extreme scarcity (and high prices) at other times. Hence, storage will become a major issue for policy makers.

The need to rapidly decarbonise the economy may impose targets and objectives that require the deployment of low-carbon technologies even at a faster rate than the market would ordinarily deliver. This may lead to additional policy-driven market interventions. Under the right regulatory environment, the next wave of technologies could be rapidly deployed at scale, as they have uses beyond the traditional power sector. While the development of batteries is being driven by and for electric vehicles, it will have important implications for both grid level and individual electricity storage.

The balance between public and private ownership and engagement within the power sector varies between countries. However, in general the most rapid move towards Phase III will occur within systems which encourage innovation and entrepreneurship, and this is likely to be delivered by the private sector, providing the market is fair and transparent and constructed to value the characteristics of renewable energies and demand-side response, and appropriate governance mechanisms are in place.

4.4.2 Business and core competencies of Phase III

In Phase III, private platforms, autonomous residents, and local grid ecosystems will co-exist with the central grid, leading to an increasing diversity of business models that range from convenient standard packages for energy services with flat rates, similar to insurances, to fully customised solutions for self-producing individuals or communities. Multinational companies, such as Google, Amazon, or Apple, will co-exist with start-ups and local initiatives.

The electricity supply industry will be forced to leave its roots as public infrastructure service and transform into truly private businesses, offering customised solutions for each consumer, while independent system operators or private transaction platforms take over responsibilities of grid control.

Digitalisation will of course be a main driver for innovation. It will cope with the complexities of the energy grid using sensors, smart meters, drones, and augmented/virtual reality in fields as diverse as predictive maintenance, customer care, and weather forecasts. Companies that embrace Artificial Intelligence and Neural Networks to detect patterns in their data will have a latent advantage over companies that solely rely on conventional computational methods. However, energy supply will always entail a technical, engineering component. Manufacturers of their own technological devices may have a competitive advantage vis-à-vis data-only companies.

Private platforms and ecosystems with micro-trading and coordination of local balancing markets will emerge. Australian start-up Power Ledger has issued the first Initial Coin Offering in the Australian energy sector and builds a blockchain-based application platform to facilitate peer-to-peer trading, monitoring of flows in the transmission grid, and plans to implement many other transaction-related functionalities.

Globalisation will allow for dispersed business operations. For example, start-ups Mobisol and SOLshare have established remote operating centres in

Berlin, from which they track and monitor the performance of their installations in sub-Saharan Africa and South Asia. Services offered in remote, rural areas integrate energy in a holistic package of services, as in the case of Solarkiosk. Energy companies and start-ups team up with players such as Total or Coca-Cola to correspond to customer needs.

In the past, trickle-down effects of innovations typically happened from industrialised to developing countries, but in Phase III new business models and digital technologies may first emerge in the developing world and then find their way to the industrialised nations. Born out of the need to experiment with autonomous micro-grids, with integrated rooftop systems, and with payment and financing methods based on the Blockchain, start-ups and entrepreneurs in developing countries may establish decentralised energy ecosystems that complement or even substitute the central grid.

This stage of transformation creates increased risks of stranded assets for existing technologies and their operators. Without transitional assistance, incumbents may delay the sector transformation. While governance bodies often focus on Phase I objectives and on how to adapt regulation, they may neglect the impact of these policies on future infrastructure.

Thinking from Phase III backwards may be an alternative approach in corporate and political decision making, allowing for a more rigorous strategy and adaptation.

4.5 References

Baumgartner, D. (2017), How to integrate a high share of renewables into the grid? HEC-ESMT Energy Course. C. Burger and J. Weinmann. Berlin, Elia Grid International.

Burger, C., Trbovich, A. & Weinmann, J. (2018), Vulnerabilities in smart meter infrastructure – can blockchain provide a solution? Results from a panel discussion at EventHorizon 2017, Background Paper, German Energy Agency DENA/ESMT. Retrieved from https://press.esmt.org/all-press-releases/blockchain-can-improve-data-security-energy-infrastructure

Cox, S. & Esterly, S. (2016), Renewable electricity standards: good practices and design considerations. Technical Report NREL/TP-6A20-65503. Golden, CO: NREL (National Renewable Energy Laboratory)

DW. (2017), *Solarworld to file for insolvency, 10 May 2017*. Retrieved February 15, 2019, from https://www.dw.com/en/germanys-solarworld-files-for-bankruptcy-again/a-43166235

Energy Brainpool. (2016), *First free-of-charge electricity flat rate in Germany, 22 September 2016*. Retrieved February 15, 2019, from https://blog.energybrainpool.com/en/first-free-of-charge-electricity-flat-rate-in-germany/

Fuchs, R. (2015). *Auf der Suche nach einem Neuanfang. Länderreport, Deutschlandfunk Kultur*. Retrieved February 15, 2019, from https://www.

deutschlandfunkkultur.de/solar-valley-in-der-region-bitterfeld-auf-der-suche-nach.1001.de.html?dram:article_id=336366

Griffith, C. (2017, July 6), Germany's sonnen offers 'free' grid power deal to solar users, *The Australian*. Retrieved February 15, 2019, from https://www.theaustralian.com.au/business/technology/germanys-sonnen-offers-free-grid-power-deal-to-solar-users/news-story/9b8f9fd7af1279f9dd49f6d226bc7a34

Lipp, J. (2007), Lessons for effective renewable electricity policy from Denmark, Germany and the United Kingdom. *Energy Policy 35*(11), 5481–5495.

NCSL. (2017), *State net metering policies.* Retrieved December 17, 2017, from http://www.ncsl.org/research/energy/net-metering-policy-overview-and-state-legislative-updates.aspx

Nicolini, M. & Tavoni, M. (2017), Are renewable energy subsidies effective? Evidence from Europe. *Renewable and Sustainable Energy Reviews, 74*,, 412–423.

Ren21. (2017a), *Renewables 2017, global status report, renewable energy policy network for the 21st century.* Retrieved from http://www.ren21.net/wp-content/uploads/2017/06/17-8399_GSR_2017_Full_Report_0621_Opt.pdf

Ren21. (2017b), *Renewables global futures report: great debates towards 100 percent renewable energy, 2017.* Retrieved from http://www.ren21.net/wp-content/uploads/2017/10/GFR-Full-Report-2017_webversion_3.pdf

Reuters. (2018), *Portugal looks to renewables as March output tops mainland power demand.* Retrieved February 15, 2019, from https://www.reuters.com/article/portugal-energy-renewables/portugal-looks-to-renewables-as-march-output-tops-mainland-power-demand-idUSL5N1RG35T

Shakoor, A., Davies, G. & Strbac, G. (2017), *Roadmap for flexibility services to 2030, Pyory, published by The Committee on Climate Change, May 2017.* Retrieved February 15, 2019, from https://www.theccc.org.uk/wp-content/uploads/2017/06/Roadmap-for-flexibility-services-to-2030-Poyry-and-Imperial-College-London.pdf

SRREN. (2012), Special report on renewable energy sources and climate change mitigation, Special Report of the International Panel on Climate Change. Intergovernmental Panel on Climate Change, Working Group III – Mitigation of Climate Change. Cambridge, UK: Cambridge University Press.

Strunz, S., Gawel, E. & Lehmann, P. (2016), The political economy of renewable energy policies in Germany and the EU. *Utilities Policy, 42*(C), 33–41.

Thalman, E. (2015, August 6), Energiewende effects on power prices, costs and industry German industry and its competitive edge in times of the Energiewende. *Clean Energy Wire.* Retrieved February 15, 2019, from https://www.cleanenergywire.org/dossiers/energiewende-effects-power-prices-costs-and-industry

van Zoonen, L. (2016), Privacy concerns in smart cities. *Government Information Quarterly, 33*(3), 472–480.

Wirth, H. (2018), *Recent facts about photovoltaics in Germany, Fraunhofer Institute, February 2018*. Retrieved February 15, 2019, from https://www.ise.fraunhofer.de/content/dam/ise/en/documents/publications/studies/recent-facts-about-photovoltaics-in-germany.pdf

Wittes, B. & Kohse, E. (2017), *The privacy paradox II: measuring the privacy benefits of privacy threats*. Retrieved February 15, 2019, from https://www.brookings.edu/wp-content/uploads/2017/01/privacy-paper.pdf

Wittes, B. & Liu, J.-C. (2015), *The privacy paradox: the privacy benefits of privacy threats*. Retrieved February 15, 2019, from https://www.brookings.edu/wp-content/uploads/2016/06/Wittes-and-Liu_Privacy-paradox_v10.pdf

CHAPTER 5

Global game changer – leading the future

The final chapter summarises the governance principles, from the country chapters, and the core competencies, drawing on the experiences of the new business model case examples, which are necessary to further accelerate the transition to a decentralised energy system.

5.1 Six reasons for decentralisation as the key driver of the global energy transformation

The energy system is undergoing radical change with decentralised renewables as a key driver due to the following trends:

(1) **The increasing competitiveness of renewable energy generation in liberalised markets – meeting grid parity and heading towards energy system parity:** renewable energies have attracted major investments in industrialised countries with an established and reliable energy system. As costs have fallen, the motivation to install solar and wind power units in these countries has shifted from publicly sponsored incentive schemes to grid parity and attractive financing models even without subsidies and government aid.

(2) **The global spread of decentralised energy generation:** since 2010, cumulative investment in distributed capacity has been around US$400 billion, moving energy supply onto rooftops and smaller acreages. Not only is this provision of services, but in countries such as Germany, Italy, or Australia community-owned initiatives have also become engaged in ownership or operation of grid infrastructure.

(3) **Decentralised storage gaining importance:** storage via batteries is a key technology to increase flexibility and adapt the energy supply system to intermittent renewable generation. Advances in storage

How to cite this book chapter:
Burger, C., Froggatt, A., Mitchell, C. and Weinmann, J. 2020. *Decentralised Energy — a Global Game Changer.* Pp. 277–283. London: Ubiquity Press. DOI: https://doi.org/10.5334/bcf.u. License: CC-BY 4.0

technologies are speeding up with the race of global car manufacturers to electrify the transport sector, and prices for lithium-ion batteries have decreased by 18 per cent between 2017 and 2018.

(4) **Decoupling growth and energy intensity via renewables and energy efficiency:** the energy intensity of the global economy is improving due to technological progress and systemic changes. Despite rising GDPs in many countries, including China, energy consumption per unit of economic output are stable or falling, also because of structural changes in their economies – with less reliance on energy-intensive industries and a shift to services and digital production.

(5) **Value creation with decentralised renewable energy generation:** renewable energies, in particular solar and wind, not only account for an increasing share of employment in the manufacturing of these technologies, but also in local value creation – either in construction and installation, or in operation and maintenance. Especially in countries in the developing world, renewable energies are drivers for local employment and value creation.

(6) **Digitalisation as an enabler of disruptive changes in energy markets:** the core competency that all industries are currently establishing is expertise in digitalisation – be it with artificial intelligence, the remote operating centres, or blockchain-based peer-to-peer platforms – digitalisation lowers barriers to entry for start-ups and facilitates new business models, increases customer choice, and is the necessary precondition for decentralised transactions.

5.2 Preparing for the three phases of the energy transformation: the 8+3+6 model

Three phases in moving from the conventional, fossil fuel-based supply structure to a decentralised, renewable system can be observed. As outlined in Section 3.1 and Chapter 4, Phase I can be associated with a niche deployment of decentralised renewable energies, contributing less than 10 per cent to total power generation. In Phase II, their contribution to total power generation amounts up to 40 per cent and becomes a major factor in the supply portfolio, whereas Phase III is characterised by decentralised renewable energies as the dominant player within a flexibly operated system with an increasing number of fully autonomous solutions not connected to a central grid.

Political and economic decision makers can prepare for the different phases of the energy transformation by thinking from Phase III (Energiewende 3.0) backwards to leapfrog or to allow the three energy phases to co-exist.

Based on the analysis of the country reports, the following eight recommendations for regulation and governance have been derived.

5.2.1 Regulation and governance: eight key principles for political decision makers

(1) **Transparency and legitimate policymaking and institutions**
The governance system needs to be able to offer clear policies and regulation that applies simultaneously to large and decentralised generation and public and private sector actors.

(2) **Availability and transparency of data**
Transparency of the system needs to enable all stakeholders to engage in the way in which the system is operated and, as the system digitalises, have access to affordable and secure data.

(3) **Customer focus, enabling customer choice**
People will affect the future energy system in three areas – as investors and operators, as willing participants and as customers who pay for innovative products that enhance their quality of life, and as voting supporters of policies and measures that deliver decarbonisation.

(4) **Markets to encourage flexibility in supply and demand**
The least expensive solution to increase flexibility and efficiency during the transformation is demand-side response. As a platform model, it does not require an expensive supply infrastructure but builds on existing assets, exploits their flexibility potential, enables peak-shaving and thus brings down peak costs.

(5) **Local system coordinators and a coexistence of the central grid and decentralised micro-grids**
Access to payments for flexibility services will be key to enable the production of power from solar and wind to be efficiently integrated into the system. This requires a more active role for regional operators of the distribution system and a greater focus on a bottom-up approach to system operation.

(6) **Including performance-based elements into sector governance**
The new regulatory framework should be based on performance rather than cost-of-service. Performance-based regulation defines desired outputs and then establishes an incentive mechanism whereby the utility or company is paid to the extent it delivers the desired outputs, as opposed to cost-of-service regulation. Inputs may change provided the desired outcomes are met, which means that there will be more flexibility of choice in delivering those outputs rather than being locked into the inputs.

(7) **Reassessing investments in the long-distance transmission grid**
If a country starts adding decentralised renewables in combination with cheaper flexibility resources, then expensive networks upgrades may not be required and reliability problems may be less relevant, thereby keeping a cap on infrastructure cost increases.

(8) **An integrative approach to sector regulation**
Ambition-driven regulation does not only ensure supply and balancing obligations are met, but wider social and environmental goals are delivered along a pre-agreed timeline. As the energy system decarbonises and decentralises, the convergence of heat, mobility and power on the distribution level requires coordinated regulatory instruments and actions. Regulators have to be flexible to establish new processes and encourage innovation across sectors.

5.2.2 Business models: three business models plus six core competencies for corporate decision makers

The real 'global game changer' related to the energy transformation is that many of the new technologies are modular and deliver services for individuals or communities at the distribution level, as well as at transmission grid scale. Electricity supply diverges from the previous model of central generating units and heads towards decentralised installations, from public ownership and large corporate entities who control the assets to a dispersed and fragmented ownership structure, often dominated by private individuals, such as homeowners, farmers, or energy associations. Corporate leaders can prepare their companies for the different phases of the energy transformation by thinking from Energiewende 3.0 backwards, preparing for the new business models via

(1) **New asset ownership models:** the infrastructure of electricity supply requires capital-intensive investments up front to establish transmission lines, generation plants and transformers, metering devices at the final users' residences, data and billing centres, and many more technical features. The rise of decentralised energy fundamentally changes ownership structures. By incentivising PV installations with feed-in tariffs, homeowners, farmers, and energy cooperatives are encouraged to install PV panels or wind turbines. Ownership of power-generating assets has become a mass-market phenomenon – both in developing and in industrialised countries. Financing these capital-intensive investments has moved from traditional methods with credits to crowdfunding, partial ownership via energy associations, and – most lately – initial coin offerings and cryptocurrencies.

(2) **New service and operating models:** business models based on efficiency services and the optimisation of the operation of energy assets – including the equipment of dwellings, such as lighting and windows – require in-depth knowledge of the complex interplay of all the energy-related components of each object. With the use of practically unlimited computing power and artificial intelligence algorithms, the barriers to entry in this field have been significantly reduced, and

new players can more quickly and easily access the market than before. Service models that include complementary convenience factors, such as safety and security features of private residencies, or allowing for assisted living, are likely to succeed in the marketplace.

(3) **New platform models:** digitalisation and the decentralised attributes of the energy transformation enhance platform models because of their low asset-intensity. Individual small-scale assets are bundled into virtual power plants to sell electricity en gross, for example on the wholesale market. Platforms are not limited to generation units. They can also integrate demand-side management for peak-shaving or ancillary grid services such as balancing energy. Aggregators open platforms and decentralised peer-to-peer trading marketplaces based on decentralised ledger technologies, such as Ethereum, and future reliability will be resolved via localised optimisation and balancing of decentralised, regional hubs.

Based on the analysis of the business cases, six core competencies for companies to develop have been identified:

(1) **Digitalisation**
All industries are currently establishing is expertise in digitalisation – a theme that is present in all interviews – be it remote operating centres, blockchain-based applications, or digital sales channels. Data management will be a key driver for commercial success; the use of Artificial Intelligence for data analytics and smart customer interaction will give a competitive edge to those companies who embrace these new machine learning tools.

(2) **Customer centricity**
At the core of the business models that trigger and accompany the transformation will be the customer. Utilities, start-ups and new entrants from other industries will offer services and value propositions that suit each customer segment. It may not be customer centricity *per se*, but rather finding the *balance* between listening to users while ensuring a high degree of standardisation. Customer centricity comes at a price, and the core competency required is to drive costs down by developing new forms of mass customisation.

(3) **Financing and enabling of asset ownership**
Companies have developed diverse strategies for how they can help prosumers to finance decentralised generation assets. Especially in developing countries, the major hurdle of a large upfront investment has been removed; in rural settings, residential owners of rooftop solar systems can generate additional revenues to repay their debts. Moreover, financial competencies stretch into the sphere of cryptocurrencies: Start-ups enable peer-to-peer transactions with virtual currencies and

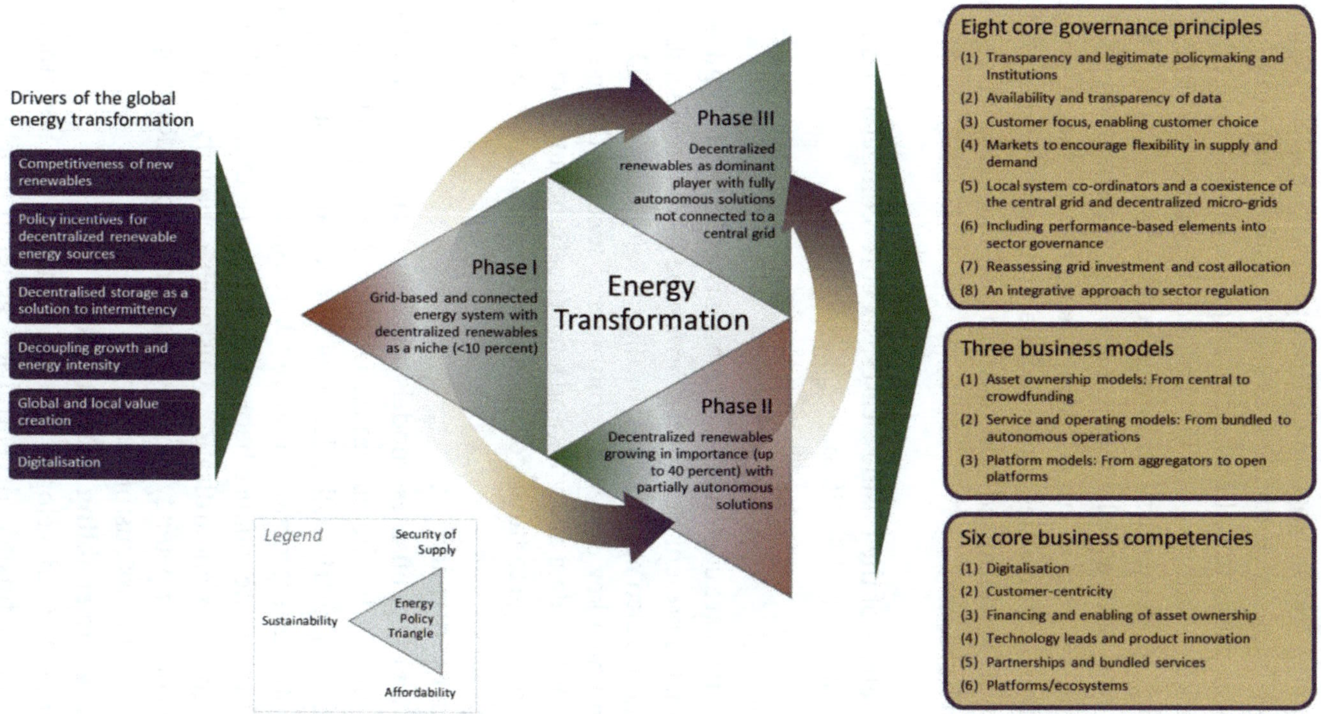

Figure 42: Governance principles and business models in the three phases of the energy transformation.
Source: Authors' contribution.

spearhead the financing revolution in terms of crowdfunding and initial coin offerings (ICOs).

(4) Technology leads and product innovation
Technology and product innovation will not only occur in the digital sphere, but also as tangible objects embedded in the new system – be it proprietary devices to steer micro-grids, customised sensors to enhance building efficiency, or drones to verify the functionality of rooftop solar panels in remote regions in Africa or South Asia. Technological advances and innovations give start-ups and established companies the leading edge and competitive differentiation. If businesses rely on pure digital service models, they can easily be crowded out by larger and financially stronger rivals.

(5) Partnerships and bundled services
New business models will also result from new partnerships that share financings, infrastructure and asset ownership. In a complex and highly dynamic market environment, no single company is able to provide all the elements of its value proposition by itself. With the increasing convergence of the energy and transport sectors, digitalisation affecting all aspects of our lives, and multinational companies entering energy markets, executives face no other option than to enter partnerships and alliances if they want to survive in the marketplace.

(6) Platforms and ecosystems
Digitalisation allows multiple players to enter markets and match supply and demand. Ownership of physical assets may not be necessary to succeed in the marketplace. The value proposition is derived instead from the coordination of providers and seekers of certain services. Sometimes these markets do not exist and have to be established, such as the market for demand response in Europe and the USA.

Figure 42 highlights the main insights.

The global transformation of the energy sector has just started. Certain major international institutions, as well as many political and corporate decision makers across all continents, are taking key roles and responsibilities in the process. If the rise of decentralised energy not only continues at the current rate of acceleration but is able to speed up as a result of good governance, then the globe may be on track for meeting the required greenhouse gas cuts whilst also benefitting from the opportunities of innovation.

CHAPTER 6

Biographies of authors

Main authors

Christoph Burger is senior lecturer at ESMT Berlin. Before joining in 2003, he worked five years in industry at Otto Versand and as vice president at the Bertelsmann Buch AG, five years at consulting practice Arthur D. Little, and five years as independent consultant focusing on private equity financing of SMEs. His research focus is in innovation/blockchain and energy markets. He is co-author of the dena/ESMT studies on 'Vulnerabilities in Smart Meter Infrastructure' and 'Blockchain in the Energy Transition', the 'ESMT Innovation Index – Electricity Supply Industry' and the book 'The Decentralized Energy Revolution – Business Strategies for a New Paradigm'.

Antony Froggatt has studied energy and environmental policy at the University of Westminster and the Science Policy Research Unit at Sussex University. He is currently an independent consultant on international energy issues and since 2007, a senior research fellow at Chatham House (also known as the Royal Institute for International Affairs). Since 2014 he has also been an honorary fellow at the Energy Policy Group at the University of Exeter. While working at Chatham House he has specialised on energy security in emerging economies with extensive work in China on the establishment and methodologies of low carbon economic development. He has also undertaken international research on public attitudes to climate change and energy security as well as to diet. He is currently working in two main areas, assessing the climate and energy policy implications of Brexit as well as evaluating the future of the electricity sector considering decarbonisation objectives and technological developments.

Catherine Mitchell is a Professor of Energy Policy at the University of Exeter, United Kingdom, and is Director of the Energy Policy Group. She has worked on energy policy issues since the 1980s. She has been a Member of numerous national and international Boards and projects. Her current area of interest is appropriate governance for innovation in energy systems. She is also a Coordinating Lead Author of the IPCC AR6 WG3 Chapter on National and Subnational Policies and Institutions.

Jens Weinmann is Program Director at ESMT Berlin. His research focuses on the analysis of strategic decision-making in corporations with respect to innovation, regulation, and competition policy, with a special interest in energy and transport. He graduated in energy engineering at the Technical University Berlin and received his PhD in Decision Sciences from London Business School. His academic experience includes fellowships at the Kennedy School of Government, Harvard University, and the Florence School of Regulation, European University Institute.

Country reports

Australia (Section 2.2):

Helen Poulter is a PhD researcher for the University of Exeter Energy Policy Group. Her PhD is part of the IGov2 – Innovation and Governance for Future Energy Systems – project. Her thesis is looking at the role that adaptive governance will play in enabling energy system transformation with a case study based on the unprecedented rise of distributed energy resources in Australia. Previously to this Helen studied a BSc Renewable Energy and worked in conservation and agriculture.

China (Section 2.3):

Liao Maolin has published more than 50 papers. She has hosted and participated in more than 40 projects, as well as several research projects on low carbon development and climate change funded by the European Union, United Kingdom, the United Nations Development Programme, and the Australian Academy of social sciences. She has participated in a number of academic works, such as the China Development Report 2012–2013, the climate change report 2011 and the green paper on the development of small and medium-sized cities, compiled by the United Nations Development Programme (UNDP) and the Institute for Urban and Environmental Studies. More than 20 papers were submitted to the office of the State Council, the central office of the central office and the Journal of the think tank. 4 of them were awarded the three awards of excellent decision-making information from the Chinese Academy of social sciences.

Dr Wei Shen is a research fellow at Institute of Development Studies. As a political economist his research interests include China's low-carbon transformation and climate change policies; China's role in global climate finance and climate governance; and South-South cooperation on climate change issues. He is particularly interested in the role of business and private actors in the process of low-carbon transformation.

Zhou Weiduo is a PhD of Graduate School of Chinese Academy of Social Science. Research assistant in Institute of Ecology and Sustainable Development, Shanghai Academy of Social Science. He studied in University of East Anglia as a visiting student from March 2017 to January 2018. He attended some Research Projects as major program of National Social Scientific Fund and major program of state condition survey of CASS. He also published 6 papers related with his major and wrote some special suggestions for the central government. His main focus is on the Sustainable Development Economics.

Denmark (Section 2.4):

Søren Djørup is an assistant professor in energy planning at Aalborg University. He holds a PhD in sustainable energy planning and has a background in economics and political economy. His research is focused on the roles of markets, regulation and policies in the transition towards renewable energy systems. He applies an interdisciplinary approach where economics and social science is combined with a technical understanding of energy systems.

Frede Hvelplund has a Dr techn. Degree in social engineering and is a Professor in Energy Planning at the Department of Planning and Development, Aalborg University, Denmark. His background is Economics and Social Anthropology. He has written many books and articles on socio-economic feasibility studies and the transition to Renewable energy systems; amongst others several 'Alternative Energy Plans' written in interdisciplinary groups together with engineers. He is a 'concrete institutional economist' and understands the market as a social construction that for decades has been conditioned to a fossil fuel and uranium based economy. Hence a transition to a 'renewable energy' economy requires fundamental changes of an array of concrete institutional rules, laws and market conditions. In December 2008 he received the EUROSOLAR European Solar Prize in Berlin.

Germany (Section 2.5):

Dr Dörte Ohlhorst has been a lecturer at the Bavarian School of Public Policy in Munich, Germany, since August 2017. Before, she was Managing Director of the Environmental Policy Research Centre at the Freie Universitaet Berlin. From 2012 to 2016, she led the project 'ENERGY TRANS' at the Environmental Policy Research Centre, which investigated different aspects of governance of the energy transition in Germany and Europe. From 2009 to 2012, she was a research associate at the German Advisory Council for the Environment and Head of the Climate and Energy Department of the Centre for Technology and Society of the Technical University of Berlin. In 2011, she founded the

Institute for Sustainable Use of Energy and Resources (iner e.V.) together with colleagues. She completed her dissertation at the Otto-Suhr-Institute for Political Science at Freie Universitaet Berlin on 'Wind Energy in Germany' in 2008. Her research focuses on energy, environmental and innovation policy, governance in the multilevel system, sustainability and participative decision-making processes as well as methods of interdisciplinary research.

India (Section 2.6):

Ranjit Bharvirkar is a Principal at the Regulatory Assistance Project where he directs RAP's India program. He has more than 17 years of experience in electricity policy analysis and technical advice and assistance to state- and national-level policymakers in the United States and India on various topics including but not limited to renewable energy, wholesale energy markets, distributed generation, energy efficiency, demand response, dynamic pricing, program evaluation, and others. In September 2018, the Oxford University Press published a volume titled 'Mapping Power: The Political Economy of Electricity in India's States' that was co-edited by Mr. Bharvirkar. He was a key contributor to India's Renewable Electricity Roadmap Initiative undertaken by the Government of India in 2014. Mr. Bharvirkar also worked at Resources for the Future (RFF) in Washington, D.C., where he was part of a team that developed a partial general equilibrium model of the US electricity sector, and analyzed the impacts of SOX, NOX, and CO2 cap-and-trade programs on electricity prices, investment and retirement of generation capacity, and distribution of benefits and costs.

Italy (Section 2.7):

Michele Gaspari is a Researcher at RSE SpA (Ricerca sul Sistema Energetico), a publicly-controlled research center: his main interests are related to innovative business models for electricity supply and to the governance and regulatory instruments that can enhance the transition towards a decentralised energy system. Michele holds a Ph.D. in Economics – Science and Management of Climate Change at Università Ca' Foscari Venezia; before joining RSE, he was Research Fellow at the Institute of Management of Scuola Superiore Sant'Anna (Pisa). He has also worked in a leasing company and in a consultancy firm active in the renewable energy sector.

Arturo Lorenzoni is Professor of Energy Economics and Electricity Market Economics in the Industrial Engineering Department of the University of Padova. He is an Electric Engineer, and holds a PhD in Energy; he is the author of more than 100 publications, mainly focusing on the economics and regulation of the power sector, as well as on the promotion of a more sustainable electricity supply. Since 1993 he cooperates with the IEFE Research Institute in Bocconi University in Milan. In 2008 Arturo co-founded the University spin-off Galileia, to offer consultancy services in the energy sector. In his career, he worked as consultant for many Italian companies and institutions. Since June 2017, he serves as Deputy Mayor in the city of Padova.

Business models

Mobisol (Section 3.7):

Klara Lindner strives to connect human-centred design with sustainable energy provision. She joined the solar company Mobisol in its infancy, led the pilot phase in East Africa, and co-developed its pioneering business model. Alongside improving Mobisol's customer experience, Klara became part of the research program Microenergy Systems in 2013, investigating service design in the bottom-of the-pyramid/energy context. As a certified Design Thinking Coach, Klara has been using various workshop settings to teach creative thinking applicable to processes of innovation and change.

www.ingramcontent.com/pod-product-compliance
Lightning Source LLC
LaVergne TN
LVHW082117200926
848484LV00008B/1968
* 9 7 8 1 0 1 3 2 9 4 8 9 1 *